中国柑橘品种

CITRUS VARIETIES IN CHINA

第二版

Second Edition

邓秀新　主编

Chief Editor DENG Xiuxin

中国农业出版社
China Agriculture Press
北　京
Beijing

图书在版编目（CIP）数据

中国柑橘品种：汉英对照/邓秀新主编．—2版．—北京：中国农业出版社，2023.8（2024.3重印）
ISBN 978-7-109-30519-9

Ⅰ．①中…　Ⅱ．①邓…　Ⅲ．①柑桔类-品种-中国-汉、英　Ⅳ．①S666.029.2

中国国家版本馆CIP数据核字（2023）第044330号

中国柑橘品种
ZHONGGUO GANJU PINZHONG

中国农业出版社出版
地址：北京市朝阳区麦子店街18号楼
邮编：100125
责任编辑：石飞华　张　利　丁瑞华
版式设计：王　晨　　责任校对：吴丽婷　　责任印制：王　宏
印刷：北京中科印刷有限公司
版次：2023年8月第2版
印次：2024年3月北京第2次印刷
发行：新华书店北京发行所
开本：889mm×1194mm　1/16
印张：18.25
字数：550千字
定价：398.00元

版权所有·侵权必究
凡购买本社图书，如有印装质量问题，我社负责调换。
服务电话：010-59195115　010-59194918

第二版编委会
The Editorial Committee for the 2nd Edition

主　　编　邓秀新
Chief Editor　　DENG Xiuxin

副 主 编　伊华林
Associate-editor　　YI Hualin

编　　委　曹　立　　陈传武　　陈克玲
　　　　　　陈善春　　陈香玲　　邓崇岭
　　　　　　邓秀新　　邓子牛　　高俊燕
　　　　　　郭文武　　黄涛江　　江　东
　　　　　　李大志　　刘建军　　米兰芳
　　　　　　吴巨勋　　徐建国　　徐　阳
　　　　　　严　翔　　曾继吾　　朱世平

Editing Members
CAO Li, CHEN Chuanwu, CHEN Keling,
CHEN Shanchun, CHEN Xiangling, DENG Chongling,
DENG Xiuxin, DENG Ziniu, GAO Junyan,
GUO Wenwu, HUANG Taojiang, JIANG Dong,
LI Dazhi, LIU Jianjun, MI Lanfang,
WU Juxun, XU Jianguo, XU Yang,
YAN Xiang, ZENG Jiwu and ZHU Shiping

英文译校　管　睿　邓秀新
English Translation and Correction
GUAN Rui and DENG Xiuxin

第一版编委会
The Editorial Committee for the 1st Edition

主　编　邓秀新
Chief Editor
DENG Xiuxin

副主编　彭成绩　陈竹生　邓子牛
　　　　　徐建国　李　健
Vice-editors
PENG Chengji, CHEN Zhusheng,
DENG Ziniu, XU Jianguo and LI Jian

编　委　蔡明段　陈竹生　邓秀新
　　　　　邓子牛　李　健　刘永忠
　　　　　彭成绩　唐小浪　徐建国
　　　　　钟广炎
Members
CAI Mingduan, CHEN Zhusheng, DENG Xiuxin,
DENG Ziniu, LI Jian, LIU Yongzhong,
PENG Chengji, TANG Xiaolang, XU Jianguo and
ZHONG Guangyan

英文翻译　管　睿　邓秀新
English Translation
GUAN Rui and DENG Xiuxin

英文审校　钟广炎
English Correction
ZHONG Guangyan

第二版前言

2008年，第十一届国际柑橘学大会在湖北武汉召开。为迎接大会的召开以及扩大我国柑橘产业在国际上的影响，自2006年我作为中国柑橘学会理事长组织国内同行编写了《中国柑橘品种》，在大会前出版。此书受到国际柑橘界的好评，也成为国内广大柑橘同行的重要参考书。十多年来，我国柑橘产业发生了巨大变化。一方面，面积和产量由当时的200多万hm^2、2 500多万t分别跃升到261.7万hm^2和4 585万t（2019年），柑橘成为我国第一大水果，而且产业效益保持稳定，为广大乡村脱贫致富提供了重要支撑；另一方面，国家加大了科技投入，2007年底建立了国家现代农业（柑橘）产业技术体系，大力研发和示范推广新品种，推进了柑橘品种多样化、供应周年化和均衡化，使我国柑橘品质得到大幅度提升，结构优化，四季有鲜果。为更好地反映这些年我国柑橘品种的变化，我组织柑橘同行对此书进行了扩充，遂成第二版。

《中国柑橘品种》第二版在原来的基础上，保持中英文介绍的风格，内容上增加了2006年以后我国审（认）定或获植物新品种权的柑橘新品种。同时，为反映产业的实际情况，将近年从境外引进推广的品种也做了介绍。为保证录入此书品种的新颖性、真实性和实践性，在遴选时考虑了下列标准：对于自主选育的品种，品种来历清楚，具有新颖性或特色，经过省、自治区、直辖市审（认）定或获得植物新品种权，且有应用价值或已有一定栽培面积；对于引进的品种，除来历比较清楚外，要求有比较大的栽培范围和面积，如沃柑和伦晚脐橙等。基于以上考虑，本次修编共增加了55个品种，其中主要是杂柑类和脐橙品种。

如果说上一版展现的是我国当时柑橘栽培品种基本样貌的话，这次增加的这些品种则代表了我国柑橘品种的研发水平。上一版我国自主选育的品种基本上以芽变和实生选种而来的品种为主，而这次展现的品种则有一批是通过杂交培育的品种，如金秋砂糖橘、金煌等，而且还有经过近40年细胞工程技术研究培育的胞质杂种华柚2号。一批新品种选育过程采用了分子标记以及组学技术，例如宗橙能明确到什么基因和DNA上的哪个碱基发生了变化。新增加的品种中，还有人工诱变的品种以及三倍体品种。分析新增品种不难看出柑橘育种的特点，即芽变仍是新品种选育的重要途径，例如从脐橙中选育了多个熟期不同的品种，椪柑中选育了无籽品种华柑4号，从琯溪蜜柚中选育了色泽变异的三红蜜柚等。这些新品种体现了我国柑橘育种技术的进步，反映出柑橘育种技术的迭代发展趋势，即从依赖自然变异到杂交创造变异，再到细胞和分子技术辅助育种。目前，我国柑橘育种同其他作物一样，朝着分子设计育种方向发展。

本书编写过程中得到了柑橘界同行的大力支持，大家踊跃提供照片并参与编写。我邀请了上一版的英文翻译管睿女士为新增的内容进行了英文翻译校核。在此，对大家的辛勤劳动表示感谢。

我相信，本书的出版能给我国乃至世界柑橘产业发展提供帮助。编写过程中难免有疏漏和不妥之处，希望读者批评指正。

邓秀新

华中农业大学教授

中国工程院院士

2022年春于北京

三峡河谷晚熟脐橙园

The navel orange plantation in the Three Gorges Valley

Preface for the 2nd Edition

In 2008, the 11th International Citrus Congress was held in Wuhan, Hubei Province. To let the participants to well understand the citrus industry and get to know the citrus varieties grown in China, I organized my colleagues to compile the book *Citrus Varieties in China* both in Chinese and in English. This book was published before the congress opening, and got participants' good praises. Also, it has provided important reference for citrus researchers and growers in China. Since then, the citrus industry in China has changed greatly; In one hand, both the acreage and total production have increased from about two million hectares, 25 million tons in 2006, to 2.617 million hectares and 45.85 million tons respectively in 2019; Now, citrus is the number one fruit crop in China based on the production quantity. The citrus industry has been keeping good profit, and the import support for alleviating the poverty and for the rural development in China. On the other hand, the government has increased the investment to the research for citrus, such as setuping China Agriculture Research System (CARS) for citrus in the end of 2007. As has greatly prompted the breeding research and new varieties demonstration as well as application to citrus industry in China; and has led to a better fresh citrus supply to the market for four seasons with improved quality. For recording the development and also for further support the citrus industry development, I organized citrus colleagues again to revise this book as the second edition.

The second edition of *Citrus Varieties in China* kept the first edition style both in Chinese and in English. It collected the new released varieties via different breeding pipelines in China since 2006. To reflect the whole picture of citrus industry, varieties introduced from abroad with large acreage or with important commercial potentiality are also included in the book. To ensure the novelty, reality and practicality of new varieties compiled in this book, we took these standards to select the variety. For the domestically bred variety, it should have a clear origin with improved traits or special characteristics, registered by the provincial governments or got Plant Variety Protection (PVP); and has been grown for certain acreage or with application potentiality in the industry. For the introduced variety from abroad, it should be grown with certain acreage in different areas, such as Orah tangor and Lane Late Navel Orange etc. Based on these, we selected 55 varieties for adding into the book. Out of them, tangor and navel orange varieties are the major.

The first edition of this book just demonstrated the global picture of citrus variety in China, whereas, the second edition presents the citrus breeding advance in China during the past decades. The first edition mainly collected citrus varieties from the chance seedling selection or from the bud sport mutation, nevertheless, this edition showed a number varieties from the cross breeding, such as Jinqiu Shatangju and Jinhuang tangor; furthermore, it includes Huayou No.2 pummelo, a cybrid from protoplast fusion with improved seedless trait, as has been researched for about 40 years. A number of the new varieties has used the molecular markers or multi-omics technology during the breeding, such as Zongcheng Navel Orange with brown peel, which is verified as a mutation of a gene controlling degration of chlorophylle since two DNA bases were changed; Also, the newly added varieties included induced mutatant and triploid respectively. From these newly bred varieties, the trend of citrus breeding can be figured out; the bud sport selection is still an efficient and practical channel for citrus breeding. This channel has generated very elite new varieties including a few early season navel orange mutants, the seedless ponkan Huagan No. 4, and the pink pummelo Sanhong miyou. In one word, these varieties have demonstrated the citrus breeding achievements in China, showing a generation-overlap development trend from utilizing natural mutation, to creating mutation via cross, and then to the cytological & molecular assisting breeding; Now, the citrus breeding in going forward to the molecular design breeding like other crops.

I would like to express great thanks to the contribution of my colleagues who provided pictures and other supports to this book. I invited Ms Guan, who translated the first edition, to joint with us again for the English correction.

I believe this new edition will benefit the citrus industry both of China and of the world. Certainly, there exist some misses or mistakes in this book; I hope the readers can give us correction and help.

DENG Xiuxin
Prof. of Huazhong Agricultural University
Member of Chinese Academy of Engineering
Spring, 2022, in Beijing

江西赣州篱壁式柑橘种植
The new plantation style (hegerow) of navel orange in Ganzhou of Jiangxi Province

第一版前言

中国是世界柑橘主要的发源地。中国柑橘已有4 000多年的栽培历史，经过长期的自然和人工选择，形成了丰富的柑橘资源宝库，现已拥有世界最丰富的柑橘资源。

中国柑橘主要分布在湖南、江西、四川、广东、福建、湖北、广西、重庆、浙江等20个省（自治区、直辖市）。2006年全国柑橘栽培面积1 814.5千hm^2，居世界首位；总产量1 789.8万t，居世界第二。主要的栽培种类有宽皮柑橘、橙、柚、杂柑及其他（枸橼、柠檬、橡檬、枳、金柑等）。

2008年国际柑橘学大会在我国武汉市召开，为了让世界同行对中国柑橘栽培的种类和品种有所了解，为了我国柑橘事业的发展，2004年中国柑橘学会决定编辑出版《中国柑橘品种》并成立了编委会。编委根据学会的要求，通力合作，广泛收集资料，认真拍摄标准图，认真撰写，于2006年完成初稿，随后分别通过各地编委审改并提出修改意见，2007年初经修改补充形成送审稿，于2007年11月由编委审稿后作了修改补充完成终审稿。

全书包括中国柑橘概况和中国柑橘栽培品种两部分。收集柑橘栽培品种（品系）168个，每个品种记述来源与分布、主要的植物学特性和果实性状、综合评价，并附标准图。这本书是比较全面地反映中国柑橘栽培品种和选育研究成果的第一本科学专著，可以作为广大柑橘从业人员和学生的参考书。

本书在编写过程中得到农业部种植业管理司和主产省（自治区、直辖市）的农业行政管理部门以及主产县的大力支持，中国主要的柑橘科研单位参与了柑橘图片的拍摄和样品提供，中国柑橘学会的广大会员为本书的出版提供了许多素材，在此一并致谢。

由于编写时间短，未能组织更广泛、更深入的调查，有的地方优良品种未能收入本书，有待以后进一步补充。此外，限于编者水平，疏漏和不妥之处在所难免，敬请读者批评指正。

中国柑橘学会
2008年1月1日

Preface for the 1st Edition

China is the major origin place of citrus in the world. With an over 4000-year history of cultivation and long-term natural and artificial selection, China has harbored the most abundant citrus germplasm in the world.

In China, citrus are mainly distributed in the 20 provinces (autonomous regions and municipalities) of Hunan, Jiangxi, Sichuan, Guangdong, Fujian, Hubei, Guangxi, Chongqing and Zhejiang etc. In 2006 the national citrus acreage reached 1 814.5 thousand hectares, which took the lead in the world, with a total annual production of 17.898 million tons, ranked the second among the citrus producing countries. The principal cultivars are Loose-skin mandarins, Oranges, Pummelos, Hybrid citrus and others (Citrons, Lemons, Limonia, Trifoliate oranges and Kumquats etc.).

In 2008, the International Citrus Congress will be held in Wuhan, China. To introduce citrus varieties grown in China to the worldwide citrus community, the Chinese Society of Citriculture decided in 2004 to organize a committee to edit and publish this book "Citrus Varieties in China". With about 3 years effort, the committee members national-widely collected materials, took photos with digital camera, selected the most typical pictures and composed the illustration to each variety. The first draft was achieved in 2006 and the delivery draft came into being at the beginning of 2007. In November, 2007, the final approved draft was accomplished after amendment and supplement.

The book consists in two parts: the general situation and the cultivated varieties in China. It introduced 168 cultivars (including strains), respectively presenting the origin and distribution, the main botanic characteristics, fruit traits and the evaluating comments. Each cultivar had a typical picture of the fruits in different views. This book in fact is the first scientific monograph that wholly introduced the citrus varieties being grown in China, as also showed the citrus breeding progress and achievement in this country. It could be used as a reference for people who work in citrus industry, and for students.

It is grateful that the compiling work has received support and assistance from the Plant Industry Administration Department, Ministry of Agriculture of the P.R.China, the agricultural administration of main citrus producing provinces (autonomous regions and municipalities) and counties. The main citrus research institutes of citrus and many members of the Chinese Society of Citriculture made great contribution to the book, including providing samples and joining in the photographing. We would like to take this opportunity to thank them all.

Since the limited time and other reasons, some local elite varieties might be missed in this book and need further supplement in the future. Criticism and rectification to the careless omission and faultiness are respected and appreciated.

The Chinese Society of Citriculture
Jan 1, 2008

中国柑橘品种　Citrus Varieties in China

目　录
Contents

第二版前言
Preface for the 2nd Edition

第一版前言
Preface for the 1st Edition

第一部分　中国柑橘概况
Part I　General Situation of Citrus Industry in China ... 1

一、中国柑橘简况　Brief Introduction ... 2

（一）中国柑橘栽培历史悠久　The History of Citriculture in China ... 2

（二）中国是柑橘重要的原产中心之一
China Is One of the Most Important Origin Centers ... 3

（三）中国柑橘资源丰富　China Is Rich in Citrus Germplasm ... 6

（四）中国是柑橘生产大国
China Is One of the Largest Citrus Producing Countries ... 7

二、中国柑橘品种结构与分布　Variety Structure and Distribution ... 10

（一）品种结构　Variety Structure ... 10

（二）分布　Distribution ... 10

第二部分　中国柑橘栽培品种
Part II　Cultivated Varieties of Citrus in China　　　　　　　　　13

一、宽皮柑橘类　Loose-skin Mandarins　　　　　　　　　14

（一）橘类　Tangerines　　　　　　　　　14

1. 八月橘　Bayueju　　　　　　　　　14
2. 本地早　Bendizao　　　　　　　　　15
　　东江本地早　Dongjiang Bendizao　　　　　　　　　16
3. 大红柑　Dahonggan　　　　　　　　　17
4. 浮廉橘　Fulianju　　　　　　　　　18
5. 桂橘1号　Guiju No.1　　　　　　　　　19
6. 行柑　Hanggan　　　　　　　　　20
7. 红橘　Hongju (Red Tangerine)　　　　　　　　　21
　　（1）大红袍　Dahongpao　　　　　　　　　22
　　（2）福橘　Fuju　　　　　　　　　23
　　（3）克里迈丁红橘　Clementine　　　　　　　　　24
　　（4）南橘　Nanju　　　　　　　　　25
8. 建柑　Jiangan　　　　　　　　　26
9. 九月橘　Jiuyueju　　　　　　　　　27
10. 马水橘　Mashuiju　　　　　　　　　28
11. 櫞橘　Manju　　　　　　　　　29
12. 明柳甜橘　Mingliutianju　　　　　　　　　30
13. 南丰蜜橘　Nanfengmiju　　　　　　　　　31
　　（1）柳城蜜橘　Liuchengmiju　　　　　　　　　32
　　（2）粤丰早橘　Yuefengzaoju　　　　　　　　　33
14. 年橘　Nianju　　　　　　　　　34
　　少核年橘　Shaohe Nianju　　　　　　　　　35
15. 椪柑　Ponkan　　　　　　　　　36
　　（1）东13椪柑　Dong 13 Ponkan　　　　　　　　　37
　　（2）鄂柑1号椪柑　E-gan No.1 Ponkan　　　　　　　　　38
　　（3）和阳2号椪柑　Heyang No.2 Ponkan　　　　　　　　　39
　　（4）华柑2号椪柑　Huagan No.2 Ponkan　　　　　　　　　40
　　（5）华柑4号椪柑　Huagan No.4 Ponkan　　　　　　　　　42
　　（6）黔阳无核椪柑　Qianyang Seedless Ponkan　　　　　　　　　43
　　（7）试18椪柑　Shi 18 Ponkan　　　　　　　　　44

（8）新生系3号椪柑　Xinshengxi No.3 Ponkan　45

（9）岩溪晚芦椪柑　Yanxiwanlu Ponkan　46

（10）永春椪柑　Yongchun Ponkan　47

（11）粤优椪柑　Yueyou Ponkan　48

（12）早蜜椪柑　Zaomi Ponkan　50

16．椪橘　Pengju　51

17．乳橘　Ruju　52

18．砂糖橘　Shatangju　53

19．酸橘　Suanju（Sour Tangerine）　55

（1）红皮酸橘　Hongpisuanju　55

（2）软枝酸橘　Ruanzhisuanju　56

20．韦尔金橘　Wilking　57

21．五月橘　Wuyueju　58

22．阳山橘　Yangshanju　59

23．早橘　Zaoju　60

24．朱橘　Zhuju　61

（1）满头红　Mantouhong　61

（2）牛肉红朱橘　Niurouhong Zhuju　62

（3）三湖红橘　Sanhuhongju　63

（4）朱红橘　Zhuhongju　64

（5）朱砂橘　Zhushaju　66

25．紫金春甜橘　Zijinchuntianju　67

（二）柑类　Mandarins　68

1．贡柑　Gonggan　68

少核贡柑　Shaohe Gonggan　69

2．黄果柑　Huangguogan　70

3．蕉柑　Jiaogan (Tankan)　71

（1）白1号蕉柑　Bai No.1 Jiaogan　72

（2）孚优选蕉柑　Fuyouxuan Jiaogan　73

（3）南3号蕉柑　Nan No.3 Jiaogan　74

（4）塔59蕉柑　Ta 59 Jiaogan　75

（5）新1号蕉柑　Xin No.1 Jiaogan　76

（6）粤丰蕉柑　Yuefeng Jiaogan　77

（7）早熟蕉柑　Zaoshu (early-ripening) Jiaogan　78

4．瓯柑　Ougan　79

5．温州蜜柑　Satsuma Mandarin　80

Ⅰ. 特早熟温州蜜柑　the Very Early-ripening　81
　　（1）大分　Oita　81
　　（2）稻叶　Inaba Wase　82
　　（3）肥之曙　Koenoakebono　83
　　（4）宫本　Miyamoto Wase　84
　　（5）国庆1号　Guoqing No.1　85
　　（6）隆园早　Longyuanzao　86
　　（7）日南1号　Nichinan No.1 Wase　87
　　（8）市文　Ichifumi Wase　88
　　（9）由良　Yura Wase　89
Ⅱ. 早熟温州蜜柑　the Early-ripening　90
　　（1）鄂柑2号　E-gan No.2　90
　　（2）宫川　Miyagawa Wase　91
　　（3）立间　Tachima Wase　92
　　（4）兴津　Okitsu Wase　93
Ⅲ. 中熟与晚熟温州蜜柑　the Middle- and Late-ripening　94
　　（1）涟红　Lianhong　94
　　（2）南柑20　Nankan No.20　95
　　（3）尾张　Owari　96

（三）杂柑类　Hybrids　97
　1. 1232橘橙　Tangor No.1232　97
　2. 不知火　Shiranui　98
　3. 春香　Hiruka Tangor　99
　4. 大雅柑　Dayagan Tangor　100
　5. 甘平　Kanpei　101
　6. 高橙　Gaocheng　102
　7. 红锦　Hongjin Tangor　103
　8. 红美人　Beni Madonna　105
　9. 红柿柑　Hongshigan　106
　10. 红玉柑　Hongyugan　107
　11. 红韵香柑　Hongyun Xianggan Tangor　108
　12. 金春　Jinchun Tangor　109
　13. 金煌　Jinhuang Tangor　111
　14. 金乐柑　Jinlegan Tangor　112
　15. 金秋砂糖橘　Jinqiu Shatangju　113
　16. 凯旋柑　Kaixuangan　114

17．明日见　Asumi　115

18．默科特　Murcott　116

19．南香　Nankou　117

20．诺瓦　Nova　118

21．清见　Kiyomi　119

22．晴姬　Harehime Tangor　121

23．秋辉　Fallglo　122

24．天草　Amakusa　123

25．沃柑　Orah　125

26．阳光1号橘柚　Yangguang No.1 Tangelo　127

27．伊予柑　Iyokan　128

28．媛小春　Himekoharu Tangor　129

29．粤英甜橘　Yueyingtianju　130

二、甜橙　Sweet Oranges　131

（一）普通甜橙　Common Sweet Orange　131

1．奥林达夏橙　Olinda Valencia Orange　131

2．冰糖橙　Bingtangcheng　132

3．长叶晚橙　Changye Wancheng　133

4．长叶香橙　Changye Xiangcheng　134

5．大红甜橙　Dahong Sweet Orange　135

6．德塔夏橙　Delta Valencia Orange　136

7．伏令夏橙　Valencia Orange　137

8．福罗斯特夏橙　Frost Valencia Orange　138

9．改良橙　Gailiang Orange　139

10．桂橙1号　Guicheng No.1　141

11．桂夏橙1号　Guixiacheng No.1 Summer Orange　142

12．哈姆林甜橙　Hamlin Sweet Orange　143

13．化州橙　Huazhou Orange　144

14．锦橙　Jincheng　145

15．锦红冰糖橙　Jinhong Bingtangcheng　146

16．锦秀冰糖橙　Jinxiu Bingtangcheng　147

17．橘湘珑冰糖橙　Juxianglong Bingtangcheng　148

18．橘湘元糖橙　Juxiangyuan Sweet Orange　149

19．卡特夏橙　Cutter Valencia Orange　150

20．柳橙　Liucheng　151

（1）暗柳橙　Anliucheng　152

　　　　（2）丰彩暗柳橙　Fengcai Anliucheng　153

　　　　（3）明柳橙　Mingliucheng　154

　21．露德红夏橙　Rohde Red Valencia Orange　155

　22．蜜奈夏橙　Midknight Valencia Orange　156

　23．糖橙　Tang Orange　157

　24．桃叶橙　Taoye Orange　158

　25．先锋橙　Xianfengcheng　159

　26．香水橙　Xiangshui Orange　160

　27．新会甜橙　Xinhui Sweet Orange　161

　28．溆浦长形甜橙　Xupuchangxing Sweet Orange　162

　29．雪柑　Xuegan　163

　　　　（1）大雪柑　Daxuegan　164

　　　　（2）零号雪柑　Linghaoxuegan　165

　　　　（3）小雪柑　Xiaoxuegan　166

　30．中柑蜜橙　Zhonggan Micheng　167

（二）脐橙　Navel Oranges　168

　1．大三岛脐橙　Omishima Navel Orange　168

　2．奉节72-1脐橙　Fengjie 72-1 Navel Orange　169

　3．奉晚脐橙　Fengwan Navel Orange　170

　4．福本脐橙　Fukumoto Navel Orange　171

　5．福罗斯特脐橙　Frost Navel Orange　172

　6．赣南1号脐橙　Ganan No.1 Navel Orange　173

　7．赣南早脐橙　Gannan Zao Navel Orange　175

　8．红肉脐橙　Cara Cara Navel Orange　176

　9．华盛顿脐橙　Washington Navel Orange　177

　10．崀丰脐橙　Langfeng Navel Orange　178

　11．伦晚脐橙　Lane Late Navel Orange　179

　12．罗伯逊脐橙　Robertson Navel Orange　180

　13．梦脐橙　Dream Navel Orange　181

　14．奈维林娜脐橙　Navelina Navel Orange　182

　15．纽荷尔脐橙　Newhall Navel Orange　184

　16．朋娜脐橙　Skaggs Bonanza Navel Orange　185

　17．黔阳冰糖脐橙　Qianyang Bingtang Navel Orange　186

　18．青秋脐橙　Qingqiu Navel Orange　187

　19．清家脐橙　Seike Navel Orange　188

　20．园丰脐橙　Yuanfeng Navel Orange　190

21．早红脐橙　Zaohong Navel Orange　192

22．宗橙脐橙　Zongcheng Navel Orange　193

（三）血橙　Blood Oranges　195

1．红玉血橙　Ruby Blood Orange　195

2．脐血橙　Washington Sanguine Blood Orange　196

3．塔罗科血橙　Tarrocco Blood Orange　197

三、酸橙　Sour Oranges　198

1．代代　Daidai　198

2．枸头橙　Goutoucheng　199

3．狮头橘　Shitouju　200

4．朱栾　Zhuluan　201

四、柚/葡萄柚　Pummelos/Grapefruits　202

（一）柚　Pummelos　202

1．HB柚　HB Pummelo　202

2．安江香柚　Anjiangxiangyou　203

3．处红柚　Chuhongyou　204

4．垫江柚　Dianjiangyou　205

5．东试早柚　Dongshizao Pummelo　207

6．度尾文旦柚　Duweiwendanyou　208

7．翡翠柚　Feicuiyou　209

8．琯溪蜜柚　Guanximiyou　210

9．桂柚1号　Guiyou No.1 Pummelo　212

10．红宝石柚　Hongbaoshi (Ruby) Pummelo　213

11．华柚2号　Huayou No.2 Pummelo　215

12．金兰柚　Jinlanyou　216

13．金香柚　Jinxiangyou　217

14．橘红　Juhong　218

15．梁平柚　Liangpingyou　219

16．坪山柚　Pingshanyou　220

17．强德勒柚　Chandler Pummelo　221

18．三红蜜柚　Sanhongmiyou　222

19．桑麻柚　Sangmayou　223

20．沙田柚　Shatianyou　225

21．四季抛　Sijipao　226

22．酸柚　Sour Pummelo　227

23. 晚白柚　Wanbaiyou　228

24. 玉环柚　Yuhuanyou　229

25. 早香柚　Zaoxiangyou　230

26. 早玉文旦　Zaoyuwendanyou Pummelo　231

（二）葡萄柚　Grapefruits　232

1. 奥兰布兰科柚　Oroblanco　232

2. 菠萝香柚　Boluoxiangyou　233

3. 胡柚　Huyou　234

4. 鸡尾葡萄柚　Cocktail Grapefruit　236

5. 星路比葡萄柚　Star Ruby　237

五、枸橼类　Citrons　238

1. 佛手　Fingered Citron　238

2. 枸橼　Citron　240

六、柠檬/檬檬　Lemons/ Limonias　241

1. 白檬檬　White Limonia　241

2. 北京柠檬　Meyer Lemon　242

3. 粗柠檬　Rough Lemon　243

4. 红檬檬　Red Limonia　244

5. 尤力克柠檬　Eureka Lemon　245

6. 云柠1号柠檬　Yunning No.1 Lemon　247

七、枳以及枳和柑橘属的杂种

Trifoliate Oranges and Its Hybrids with Citrus Genus　248

1. 枳　Trifoliate Orange　248

2. 枳橙　Citrange　250

3. 枳柚　Citrumelo　251

八、金柑　Kumquats　252

1. 长寿金柑　Changshou Kumquat　252

2. 脆蜜金柑　Cuimi Kumquat　254

3. 桂金柑1号　Guijingan No.1 Kumquat　256

4. 桂金柑2号　Guijingan No.2 Kumquat　258

5. 金弹　Jindan　260

6. 金豆　Jindou　261

7. 罗浮　Luofu　262

8. 罗纹　Luowen　263

9. 四季橘　Sijiju　265

主要参考文献
Main References   266

中国柑橘品种中文名称索引
Index to Citrus Varieties in China (in Chinese)   267

中国柑橘品种英文名称索引
Index to Citrus Varieties in China (in English)   270

第一部分
中国柑橘概况

Part I General Situation of Citrus Industry in China

一、中国柑橘简况
Brief Introduction

(一) 中国柑橘栽培历史悠久
The History of Citriculture in China

中国是世界上栽培柑橘历史最早的国家，迄今已有四千多年的历史。《禹贡》有"厥包橘柚锡贡"记载，说明在夏禹时代（约公元前21世纪），已有橘、柚（香橙）、枳（酸橙）的栽培，并把橘和柚作贡品。春秋战国时期（公元前770—前221年）记载"果之美者有江浦之橘，云梦之柚"，"蜀汉、江陵千树橘。"两千多年前屈原写的《离骚》中有《橘颂》一章。说明二三千年前，四川、重庆、湖北、湖南一带，柑橘已盛行栽培。世界第一部柑橘专著，宋代韩彦直所著《橘录》（1178年），记载了浙江温州有27种柑橘，介绍繁殖技术、栽培技术、病虫防治、采收、贮藏和加工等技术。充分说明中国柑橘栽培有悠久历史。

As a pioneer in growing citrus, China has a more than 4000-year long history of citrus production. It was recorded in *Yugong* written in the time of Xiayu (about 21st century B.C.) that Ju (Mandarin), You (Junos) and Zhi (Sour orange) were listed as tribute items, which indicated that citrus fruits were cultivated and used as tributes at that time. In Spring-Autumn and Warring-States Periods (from 770 B.C. to 221 B.C.) it was literarily described that "the most delicious fruits are 'Ju'(Mandarins) from Jiangpu and 'You'(Poumeloes) from Yunmeng" and that "there are thousands of trees of 'Ju' in Shuhan and Jiangling". There is one chapter called "Ju Song" or " The Song of Mandarin" in the poem of *Li Sao* compiled 2 000 years ago by the famous poet Qu Yuan, indicating that citrus was widely cultivated in Sichuan, Chongqing, Hubei and Hunan at that time. The world's first monograph of citrus, *Ju Lu*, written by Han Yanzhi in the Song Dynasty (1178 A.D.), recorded 27 citrus cultivars from Wenzhou, Zhejiang Province, together with technical description of citrus propagation, cultivation, disease and pest control, harvesting, storage and processing.

（二）中国是柑橘重要的原产中心之一
China Is One of the Most Important Origin Centers

中国是柑橘重要的原产中心之一，被誉称为世界柑橘资源的宝库。枳、金柑、宽皮柑橘、橙、柚原产我国。枳原产长江上游，广泛分布于西南和中南各省（自治区、直辖市），陕西、甘肃、山西、山东、河南等省也有分布。广东、广西、福建、浙江、湖南、江西等省（自治区）均有野生的山金柑（金豆）。金沙江、大渡河上游的河谷地带发现有野生、半野生的黄果（甜橙）、白橘（椪柑）、柚和大翼橙的红河橙、马蜂柑等原始群落。大巴山、武当山、巴东三峡、广西龙胜山、苗儿山、云贵高原发现有野生的宜昌橙。湖南道县、莽山和江西崇义发现野生橘，广东龙门县南昆山发现三百多年生的野生橘柚自然杂交种（当地人称香橙）以及近百年生的野生金柑。

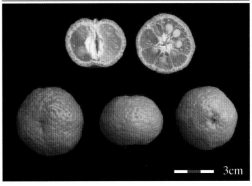
道县野橘 Daoxian wild tangerine

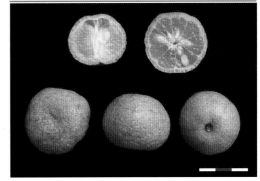
莽山野橘 Mangshan wild tangerine

As one of the most important origin centers, China is well known as the treasury of citrus resources of the world. Trifoliate orange, Kumquats, Loose-skin Mandarins, Oranges and Pummelos originated in China. Trifoliate orange, originated in the upper reaches of the Yangtze River, is widely distributed in southwest and central-south China, and is also spotted in other provinces such as Shaanxi, Gansu, Shanxi, Shandong and Henan etc. Wild Shanjingan (*Fortunella hindsii*) grows in Guangdong, Guangxi, Fujian, Zhejiang, Hunan and Jiangxi provinces. In the upper valley regions of the Jinshajiang River and the Daduhe River there are wild and semi-wild Huangguo (Sweet orange), Baiju (Ponkan), Pummelos, and the original communities of wild Honghe papeda (*Citrus hongheensis*) and Mafenggan (*Citrus hystrix*). Yichangcheng (*Citrus ichangensis*) has been discovered in Daba Mountain, Wudang Mountain, the Three Gorges area of Badong County, Longsheng Mountain and Miaoer Mountain of Guangxi, and Yungui plateau. In recent years some wild tangerine have been discovered in Daoxian and Mangshan of Hunan and in Chongyi of Jiangxi. A 300 years old natural tangelo tree and a more than 100 years old wild kumquat tree have been discovered in Nankun Mountain of Longmen County, Guangdong Province.

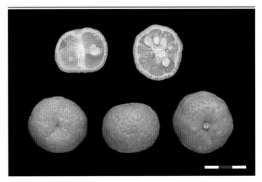
崇义野橘 Chongyi wild tangerine

崇义野橘 Chongyi wild tangerine

野生橘柚 Wild tangelo

野生金豆 Wild jindou

野生小枸橼 Wild small citron

第一部分 中国柑橘概况
Part I General Situation of Citrus Industry in China

山金柑 The Hongkong kumquat

野生柑橘类型，广泛分布于南岭山脉等地，近年发掘到单胚类型，播种当年开花结果，是理想的柑橘功能基因研究材料
A wild type of citrus, is widely distrubuted in the Nanling Mountains area. During past years, the monoembryogenic type of this species has been found and developed as a model plant for the function genomic research for citrus.

山金柑与血橙杂交后代果实
The fruits of the progeny from the cross of Hongkong kumquat with blood orange

(三) 中国柑橘资源丰富 China Is Rich in Citrus Germplasm

中国柑橘资源非常丰富，国家柑橘资源圃保存1 200多份资源。在生产上栽培种主要有柑橘属以及少量的金柑属、枳属。柑橘属由于长期的栽培和选育，产生了许多天然的实生变异和芽条变异类型的新品种、品系。21世纪以来，我国柑橘杂交育种工作得到加强，通过杂交育种培育出金秋砂糖橘、金煌等杂柑新品种，通过原生质体融合培育出无籽胞质杂种华柚2号。各省（自治区、直辖市）还先后从国外引进了不少新的品种，丰富了我国的柑橘资源。

China is abundant in citrus resources. More than 1 200 accessions of germplasm have been collected in the National Citrus Germplasm Repository. The majority of the cultivated cultivars are from the genus *Citrus* and a few from *Fortunella* and *Poncirus*. Many new cultivars from natural mutations of seedlings and budwoods have been developed by long-term cultivation and selection. The crossing breeding has been strengthened since the new century; new varieties such as Jinqiu Shatangju and Jinhuang tangor via crossing were released during the past decade; New pipeline via protoplast fusion also began to generate seedless cybrids such as Huayou No.2 pummelo. In addition, new varieties lines introduced from abroad have further enriched the genetic pool of citrus resources of China.

不同类型的柑橘果实 Different citrus types with varied fruit size and shape

（四）中国是柑橘生产大国
China Is One of the Largest Citrus Producing Countries

中国是世界柑橘生产大国，2019年种植面积达261.73万 hm^2，产量4 584.54万t，均居世界首位（表1）。

China is the largest citrus producing country both for acreage and for the total production. In 2019, China's citrus acreage reached 2 617.3 thousand hectares, with a total production 45.85 million tons (Table 1).

表1　中国柑橘2019年种植面积与产量
Table 1　China Citrus Acreage and Yield in 2019

省（自治区、直辖市）Province (autonomus region and municipality)		面积（khm^2）Acreage (thousand hectares)	产量（kt）Yield (thousand metric tons)
安徽	Anhui	2.2	30.8
重庆	Chongqing	221.7	2 950.7
福建	Fujian	138.5	3 657.6
甘肃	Gansu	0.2	1.5
广东	Guangdong	236.3	4 648.0
广西	Guangxi	438.5	11 245.2
贵州	Guizhou	72.8	523.7
海南	Hannan	8.3	84.7
河南	Henan	4.5	46.3
湖北	Hubei	232.8	4 782.2
湖南	Hunan	399.9	5 604.7
江苏	Jiangsu	2.2	28.7
江西	Jiangxi	336.0	4 131.8
陕西	Shaanxi	23.7	503.6
上海	Shanghai	3.5	108.4
四川	Sichuan	323.1	4 577.3
西藏	Tibet	0.2	0.6
云南	Yunnan	84.3	1 085.7
浙江	Zhejiang	88.6	1 834.0
合计	Total	2617.3	45 845.5

注：表1未统计台湾省的产量和面积。
Note：Table 1 did not include the data of Taiwan Province.

江西赣南脐橙园　Navel orange orchard in Gannan of Jiangxi Province

福建琯溪蜜柚园　Guanximiyou orchard in Pinghe of Fujian Province

湖北当阳柑橘园　Citrus orchard in Dangyang of Hubei Province

第一部分　中国柑橘概况
Part I General Situation of Citrus Industry in China

重庆忠县甜橙园
Sweet orange orchard in Zhongxian of Chongqing Municipality

湖南道县脐橙园
Navel orange orchard in Daoxian of Hunan Province

广东杨氏柑橘园　Citrus orchard of Guangdong Province

二、中国柑橘品种结构与分布
Variety Structure and Distribution

（一）品种结构　Variety Structure

中国柑橘栽培品种主要有宽皮柑橘、橙、柚、杂柑，其他包括枸橼、柠檬、橡檬、枳、金柑。其中宽皮柑橘的产量占柑橘总产量的58%，橙占24%，柚占11%，柠檬占4%，其他3%。

The main cultivars of citrus in China are Loose-skin mandarins, Sweet oranges, Pommeloes and Hybrid varieties. The others are Citrons, Lemens, Limes, Trifoliate oranges and Kumquats. The Loose-skin mandarins account for 58 % of the total yield, Sweet oranges for 24%, Pummelos for 11%, lemens for 4% and others for 3% respectively.

（二）分布　Distribution

中国柑橘主要分布在亚热带地区。其中南亚热带地区包括广东、福建、台湾三省大部分地区及云南西双版纳、普洱（原思茅）、文山、红河州等地；年均温22℃左右，冷月均温10℃以上，最低温度一般在0℃以上，当寒流侵袭时可降到0℃以下，但持续时间极短，≥10℃的积温6 500～8 000℃；年降水量1 200～2 200mm；以生产甜橙、宽皮柑橘、柚为主。中亚热带包括浙江温州地区、湖南道县、江西赣州、广西桂林以南，以及米仓山、大巴山以南的四川盆地，湖北宜昌以西的长江峡谷地区，广东的韶关北部；年均温17～21℃，1月均温7～9℃，最低气温一般在-3℃以上，≥10℃的积温5 500～6 500℃；年降水量1 000mm以上，东南沿海可达2 000mm；以生产宽皮柑橘、甜橙、柚为主，是我国最重要的柑橘产区。北亚热带包括浙江、湖北、江西、湖南等省的大部分地区，陕西、甘肃、河南、安徽、江苏南部部分地区和上海市的长兴岛；年均温15～17℃，1月均温5～7℃，最低气温一般-7～-5℃，≥10℃的积温4 500～5 500℃；年降水量1 000～1 500mm；以生产宽皮柑橘为主。详细分布见表2。

表2 中国柑橘主栽品种与主要砧木分布
Table 2 The Distribution of Main Cultivars and Rootstocks of Chinese Citrus

省（自治区、直辖市） Province (autonomus region and municipality)	主栽种类和品种 Main types and cultivars	主要砧木 Main rootstocks
重庆 Chongqing	脐橙、夏橙、血橙、锦橙、柚、温州蜜柑、红橘、椪柑、杂柑 Navel oranges, Summer orange, Blood orange, Jincheng sweet orange, Pummelo, Satsuma mandarins, Hongju, Ponkan, Hybrid mandarins	枳、红橘、枳橙、酸柚 Trifoliate orange, Hongju, Citrange, Sour pummelo
福建 Fujian	琯溪蜜柚、葡萄柚、椪柑、脐橙、温州蜜柑、红橘、金柑 Guanxi Miyou pummelo, Grapefruit, Ponkan, Navel orange, Satsuma mandarins, Hongju, Kumquats	枳、红橘、酸柚 Trifoliate orange, Hongju, Sour pummelo
广东 Guangdong	砂糖橘、贡柑、蕉柑、马水橘、春甜橘、年橘、茶枝柑、红江橙、沙田柚、琯溪蜜柚、化橘红、柠檬 Shatangju, Gonggan, Jiaogan, Mashuiju, Chuntianju, Nianju, Chazhigan, Hongjiangcheng, Shatianyou pummelo, Guanxi Miyou pummelo, Huajuhong, lemon	酸橘、三湖红橘、红橡檬、酸柚 Sour tangerine, Sanhuhongju, Red limonia, Sour pummelo
广西 Guangxi	砂糖橘、马水橘、沃柑、茂谷柑、温州蜜柑、椪柑、沙田柚、暗柳橙、脐橙、夏橙、金柑 Shatangju, Mashuiju, Orah tangor, Murcott, Satsuma mandarins, Ponkan, Shatianyou pummelo, Anliucheng, Navel oranges, Summer orange, Kumquats	枳、资阳香橙、酸橘、酸柚橡檬 Trifoliate orange, Ziyang Xiangcheng, Sour tangerine, Sour pummelo, Limonia
贵州 Guizhou	温州蜜柑、大红袍、椪柑、琯溪蜜柚 Satsuma mandarins, Dahongpao, Ponkan, Guanxi miyou pummelo	枳 Trifoliate orange
海南 Hainan	暗柳橙、红江橙、琯溪蜜柚 Anliucheng, Hongjiangcheng, Guanximiyou pummelo	酸橘、红橡檬 Sour tangerine, Red limonia
湖北 Hubei	温州蜜柑、椪柑、脐橙、锦橙、桃叶橙、夏橙、柚、杂柑 Satsuma mandarins, Ponkan, Navel oranges, Jincheng, Taoyecheng, Summer orange, Pummelo, Hybrid Mandarins	枳、红橘、枳橙 Trifoliate orange, Hongju, Citrange
湖南 Hunan	温州蜜柑、椪柑、脐橙、冰糖橙、大红甜橙、柚 Satsuma mandarins, Ponkan, Navel oranges, Bingtangcheng, Dahongtiancheng, Pummelo	枳、酸柚 Trifoliate orange, Sour pummelo
江苏 Jiangsu	温州蜜柑、本地早 Satsuma mandarins, Bendizao	枳、朱红橘 Trifoliate orange, Zhuhongju
江西 Jiangxi	脐橙、南丰蜜橘、温州蜜柑、椪柑、柚、金柑 Navel oranges, Nanfengmiju, Satsuma mandarins, Ponkan, Pummelo, Kumquats	枳、酸柚 Trifoliate orange, Sour pummelo
上海、安徽、陕西、甘肃、河南 Shanghai, Anhui, Shaanxi, Gansu and Henan	温州蜜柑、朱红橘 Satsuma mandarins, Zhuhongju	枳 Trifoliate orange
四川 Sichuan	杂柑、温州蜜柑、红橘、脐橙、血橙、锦橙、柚、柠檬、椪柑 Hybrid mandarins, Satsuma mandarins, Hongju, Navel oranges, Blood orange, Jincheng, Pummelo, Lemon, Ponkan	枳、资阳香橙、红橘、酸柚 Trifoliate orange, Ziyang Xiangcheng, Hongju, Sour pummelo
台湾 Taiwan	椪柑、蕉柑、雪柑、柳橙、茂谷柑、柚 Ponkan, Jiaogan, Xuegan, Liucheng, Murcott, Pummelo	酸橘、红橡檬、枳、酸柚 Sour tangerine, Red limonia, Trifoliate orange, Sour pummelo
西藏 Tibet	皱皮柑、枸橼 Zhoupigan, Citron	枳 Trifoliate orange
云南 Yunnan	温州蜜柑、椪柑、沃柑、脐橙、冰糖橙、柠檬、柚、枸橼 Satsuma mandarins, Ponkan, Orah tangor, Navel oranges, Bingtangcheng, Lemon, Pummelo, Citron	枳、资阳香橙、酸柚 Trifoliate orange, Ziyang Xiangcheng, Sour Pummelo
浙江 Zhejiang	温州蜜柑、椪柑、本地早、杂柑、柚、胡柚、金柑、枸橼 Satsuma mandarins, Ponkan, Bendizao, Hybrid mandarin, Pummelo, Huyou, Kumquats, Citron	枳、枸头橙、酸柚 Trifoliate orange, Goutoucheng, Sour pummelo

Most citrus fruits in China are distributed in subtropical regions. The south subtropical regions include the most areas in Guangdong, Fujian, Taiwan provinces and Xishuangbanna, Puer, Wenshan, Honghe districts of Yunnan Province, where the annual average temperature is around 22℃, the average temperature in the coldest month is above 10℃, the lowest temperature is above 0℃ and may drop below 0℃ when cold current comes but lasts only a very short time, the annual accumulated temperature is 6 500~8 000℃ (≥10℃), and the annual precipitation is 1 200~2 200 mm. These regions are the major production areas of Sweet oranges, Loose-skin mandarins and Pummelos. The central subtropical regions include areas stretching from Wenzhou of Zhejiang, Daoxian of Hunan, Ganzhou of Jiangxi, the south of Guilin, areas of south Daba and Micang Mountains of Sichuan, and areas from the Three Gorges area of Yangtze River to the west of Yichang of Hubei, and the north of Shaoguan of Guangdong. In these areas, the average annual temperature is 17~21℃, the average temperature in January is 7~9℃, the lowest temperature is above −3℃, the annual accumulated temperature is 5 500~6 500℃ (≥10℃), and the annual precipitation is above 1 000 mm but may exceed 2 000 mm in the southeast coastal area. These areas grow mainly Loose-skin mandarins, Sweet oranges, and Pummelos, and are the most important citrus producing areas in China. The north subtropical regions include the most parts of Zhejiang, Hubei, Jiangxi, Hunan, and Changxing Island of Shanghai, where the average annual temperature is 15~17℃, the average temperature in January is 5~7℃, the lowest temperature is −7~−5℃, the annual accumulated temperature above 10℃ is 4 500~5 500℃, and the annual precipitation is 1 000~1 500mm, produce mainly Loose-skin mandarins. The detailed citrus distribution is listed in Table 2.

2003年，农业部发布了中国柑橘优势区域规划，根据气候、土壤以及交通、生产现状等因素将中国柑橘生产的优势区域划分为三带加特色基地。分别是长江上中游柑橘带、浙南、闽西、粤东柑橘带以及赣南、湘南、桂北柑橘带；特色基地包括云南早熟蜜橘、江西南丰蜜橘、广西融安金柑等。三条柑橘带各有特色，其中长江上中游柑橘带主产脐橙和夏橙等鲜食及加工品种，浙南、闽西、粤东柑橘带主产鲜食和加工的宽皮柑橘，赣南、湘南、桂北柑橘带主产鲜食甜橙。近十年来，还形成了北起湖北丹江、南到湖南邵阳一条新的宽皮柑橘优势带；以及西江流域晚熟宽皮柑橘优势带，即从云南的玉溪沿着西江到广东肇庆，主要栽培的是砂糖橘、沃柑等晚熟柑橘，这两条带弥补了东部宽皮柑橘带的不足。

In 2003, the Ministry of Agriculture of China promulgated the layout of the citrus advantage regions in China, which consists of three geological belts plus featured production bases based on climate, soil, transportation conditions and current production situations. The three belts are referred as the belt of the middle and upper Yangtze River region, the belt of southern Zhejiang, western Fujian and eastern Guangdong, and the belt of southern Jiangxi, southern Hunan and northern Guangxi. The featured production bases include the early-maturing Satsuma in Yunnan, the Nanfengmiju in Jiangxi and the Rong-an kumquat in Guangxi, etc. The three citrus producing belts have their own features. The middle and upper Yangtze River belt produces Naval oranges for fresh market and Valencia oranges for juice-processing industry. The southern Zhejiang, western Fujian and eastern Guangdong belt is the major producer of fresh and processed mandarins. The third belt grows mainly fresh Sweet oranges. During the past decade, two new mandarin producing belts have been formed, one is from Danjiang of Hubei Province to Shaoyang of Hunan Province ; another is Xijiang River citrus producing belt, begening from Yuxi of Yunnan Province, along the Xijiang River, and end in Zhaoqing of Guangdong, where the late season loose skin mandarin including Shatangju tangerine and Orah tangor are produced. These lately formed two belts play important role of supplement to the insufficient production of the east mandarin producing belt.

第二部分
中国柑橘栽培品种

Part II Cultivated Varieties of Citrus in China

中国柑橘品种 Citrus Varieties in China

一、宽皮柑橘类
Loose-skin Mandarins

（一）橘类 Tangerines

1. 八月橘

来源与分布：八月橘原产广东四会，四会等地有少量栽培，广西有引种。

主要性状：树势较强壮。春梢叶片大小为7.2～9.1cm×3.8～4.6cm（长×宽），椭圆形。果实扁圆形，橙黄色，大小为5.4cm×3.8cm（横×纵），果顶平，微凹，果皮容易剥离。可溶性固形物（TSS）10.0%～11.5%，酸0.6%，果肉清甜，较不化渣。种子16.6粒/果。成熟期10月下旬。早结丰产性好，果实比砂糖橘大，耐贮性稍差。主要砧木为酸橘等。

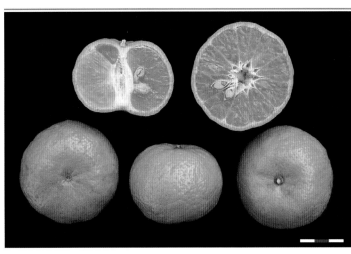

1. Bayueju

Origin and Distribution: Bayueju, originated in Sihui, Guangdong Province; is now grown on a small scale in Sihui. Guangxi had introduced it.

Main Characters: Tree relatively vigorous. Leaf size 7.2~9.1cm (length) × 3.8~4.6cm (width), elliptic. Fruit obloid, orange-yellow, size 5.4cm (diameter) × 3.8cm (heigth), apex truncate, slightly depressed, easily peeling. TSS 10.0%~11.5%, acid content 0.6%; pulp fresh sweet, not very melting. 16.6 seeds per fruit. Matures in late October. The cultivar is early fruiting and very productive. Fruit size is larger than Shatangju. The storage quality of fruits is less good. The main rootstock used is Sour tangerine.

2. 本地早

来源与分布：本地早又名天台山蜜橘。原产浙江黄岩。主产浙江、江西，四川、广东、广西、云南、福建等地有少量栽培。

主要性状：树势强，树冠圆头形，分枝多而细密，枝细软。春梢叶片大小为7.0～9.0cm×3.0～4.0cm，长椭圆形，叶缘锯齿明显。果实扁圆形，橙黄色，大小为5.5 cm×4.5cm，果顶微凹，皮略显粗糙，易剥离。TSS 11.0%～13.0%，酸0.5%～0.7%，种子2.4粒/果。成熟期11月上中旬。该品种味甜酸少，有香气、化渣，品质优良。是鲜食和制罐兼优品种。丰产耐寒，果实不耐贮藏。平地缓坡山地用枳、海涂用枸头橙作砧木。

2. Bendizao

Origin and Distribution: Bendizao or Tiantaishanmiju is originated in Huangyan, Zhejiang Province, and mainly grown in Zhejiang and Jiangxi. There is small-scale cultivation in Sichuan, Guangdong, Guangxi, Yunnan and Fujian.

Main Characters: Tree vigorous with spheroid crown; branches delicate and dense. Leaf size 7.0~9.0cm×3.0~4.0cm, long elliptic, margin dentate. Fruit obloid, orange-yellow, size 5.5cm×4.5cm, apex slightly depressed, rind slightly rough, easily peeling. TSS 11.0%~13.0%, acid content 0.5%~0.7%, 2.4 seeds per fruit. Matures in early to middle November. Flesh sweet, low in acid content, fragrant and melting. The cultivar is excellent for both fresh and canned consumption. It is productive and cold-tolerant although not so tolerant to storage. The rootstocks used are Trifoliate orange in orchards on flat land, slow-sloped and terraced hills, and Goutoucheng in costal areas.

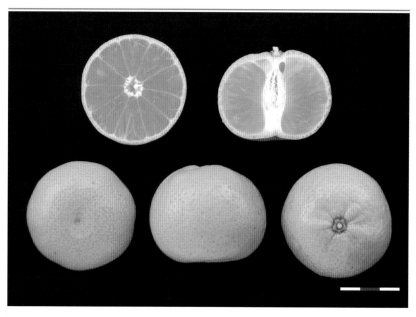

东江本地早 东江本地早为普通本地早少核良种芽变，2007年通过浙江省非主要农作物良种认定，主产浙江台州。

树势略强，树冠圆头形，分枝多而细密，枝细软。果实扁圆形，橙黄色，单果重58～75g，果皮橙色、较薄、易剥离，果面光滑，汁液多，化渣性好，TSS 12.3%，酸0.5%，可食率79%，少核或无核，种子0.6粒/果。11月上旬成熟，比普通本地早提早7d成熟，丰产耐寒，果实货架期短。适宜中、北亚热带，年光照时数在2 000h左右的柑橘产区。

Dongjiang Bendizao This variety is a less-seedy bud sport selection from the common Bendizao tangerine. It was registered as an elite variety of non-major crops in Zhejiang Province in 2007. It is mainly produced in Taizhou, Zhejiang.

Tree considerably vigorous with spheroid crown; branches delicate and dense. Fruit obloid, orange-yellow, weight 58～75g; pericarp orange, thin and smooth, easy to peel. Pulp juicy and melting, TSS 12.3%, TA 0.5%, edible portion 79%. Few seeds or seedless with an average of 0.6 seeds per fruit. Mature in early November, 7 days earlier than the original Bendizao tangerine. The cultivar is productive and cold-tolerant, although the fruits have a short shelf life. It is suitable for the citrus-producing areas where the annual sunlight hours are about 2 000h in the central and north subtropical regions.

3. 大红柑

来源与分布：大红柑又名茶枝柑，原产广东新会，主产广东江门、新会、台山、恩平、开平，英德、惠州有少量栽培，广西、浙江等地有引种。

主要性状：长势中等，树冠半圆头形，枝条细长密集，有短小细刺。春梢叶片大小为5.8～6.8cm×2.5～2.9cm，长椭圆形，叶缘波状，翼叶小。花大小介于橙、橘之间，为完全花。果实扁圆形，橙黄色，大小为5.5～6.0cm×4.0～4.5cm，果顶平，微凹，果蒂部有放射沟纹，果皮易剥离。TSS 10.5%～12.5%，酸0.5%～0.7%。种子10粒/果，成熟期11月中旬至12月上旬。该品种丰产性好，果皮是制中药陈皮及其系列产品的正宗原料。主要砧木为酸橘、三湖红橘。

3. Dahonggan

Origin and Distribution: Dahonggan, or Chazhigan, originated in Xinhui, Guangdong, and is mainly grown in Jiangmen, Xinhui, Taishan, Enping, Kaiping. There is small-scale distribution in Yingde, Huizhou and introduction in Guangxi and Zhejiang etc.

Main Characters: Tree semi-vigorous, half-spheroid, branches long, slender and dense, with small and short spines. Leaf size 5.8~6.8cm×2.5~2.9cm, long-elliptic, margin sinuate, petiole wing small. Complete flower, size between orange and mandarin. Fruit obloid, yellowish orange, size 5.5~6.0cm×4.0~4.5cm; apex truncate, slightly depressed; base striped with radial furrows, easily peeling. TSS 10.5%~12.5%, acid content 0.5%~0.7%, 10 seeds per fruit. Maturity period ranges from mid-November to early December. The variety is very productive. Rind is the genuine material for the production of Chinese medicine Chenpi and Chenpi derivatives. The main rootstocks used are Sour tangerine and Sanhuhongju.

4. 浮廉橘

来源与分布：浮廉橘又名书田橘，原产粤东山区，主产广东的河源、惠州等市。

主要性状：树势健壮，树冠圆头形，枝细密。春梢叶片大小为6.7～9.6cm×2.2～3.2cm，披针形或长椭圆形，叶尖渐尖有小凹口，叶缘锯齿浅，微波浪状，翼叶小，线形。果实扁圆形，橙黄色，大小为5.5～6.0cm×4.2～5.0cm，果顶平微凹。TSS 10.0%～12.0%，酸0.8%～1.0%，味甜酸，品质中等。种子15～20粒/果。成熟期11月中下旬。该品种适应性强，河源市紫金县仍有百年生老树。

4. Fulianju

Origin and Distribution: Fulianju, or Shutianju, originated in the mountainous area of east Guangdong Province, is mainly grown in Heyuan, Huizhou of Guangdong.

Main Characters: Tree vigorous with spheroid crown, branches slender and dense. Leaf size 6.7~9.6cm×2.2~3.2cm, lanceolate or long elliptic, apex acute and notched, margin shallowly dentate or slightly sinuate; petiole wing linear and small. Fruit obloid, orange-yellow, 5.5~6.0cm×4.2~5.0cm; apex truncate, slightly depressed. TSS 10.0%~12.0%, acid content 0.8%~1.0%, taste sweet and sour, of medium quality. 15~20 seeds per fruit. Matures from middle to late November. The variety has strong adaptability. There exist large Fulianju trees older than 100 years in Zijin County, Heyuan, Guangdong.

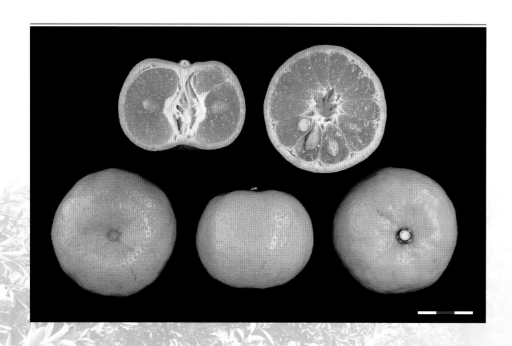

5. 桂橘1号

来源与分布：桂橘1号是从蜜广橘芽变中选育的早熟柑橘新品种，在广西桂林市兴安县界首镇城东村1974年种植的南丰蜜橘果园中发现。2014年通过广西农作物品种审定委员会审定。

主要性状：果实扁圆形，果皮橙黄色，果面光滑，有光泽，果肉橙黄色，化渣，味清甜；单果重46.8g，最大单果重63.3g，果形指数0.72；种子0.9粒/果，单胚，子叶淡绿色。可食率78.0%，果汁率61.3%，TSS12.3%，全糖11.0%，酸0.6%，每100ml果汁维生素C含量21.2mg，品质上等。果实生育期160d，在桂林兴安10月上中旬成熟。桂北地区及柳州都能栽培，适宜在南丰蜜橘种植的区域推广。

5. Guiju No.1

Origin and Distribution: Guiju No.1 was selected from a bud mutation of Miguangju in 1974 in Jieshou, Xing'an County, Guangxi, registered in Guangxi in 2014.

Main Characters: Guiju No.1 tangerine is medium-vigorous with spheroid canopy. Fruit oblate, rind orange-yellow, smooth and glossy. Pulp orange-yellow, taste pure sweet. Fruit weight 46.8g (maximum 63.3g), fruit shape index 0.72; Seeds 0.9 per fruit, mono-embryonic, cotyledon light green. Edible portion 78.0%, juicing rate 61.3%, TSS 12.3%, total sugar 11.0%, TA 0.6%, Vitamin C 21.2mg/100ml. Quality excellent. Fruit growth period 160 days, maturity in the early-to-mid October in Xing'an County, Guangxi. It is suitable for Northern Guangxi and Liuzhou, Guangxi; Production area for Nanfengmiju.

6. 行柑

来源与分布： 行柑又名鱼冻柑、四会柑，原产广东四会，在肇庆有少量栽培，浙江、湖北等省有引种。

主要性状： 树势中等，树冠圆头形，枝条细密。春梢叶片大小为 5.7～7.7cm×2.9～3.8cm，广卵圆形，叶缘锯齿浅，翼叶小。花小，为完全花。果实扁圆形，橙黄色，大小为6.6～7.6cm×5.1～5.3cm，果顶平，微凹，果皮容易剥离。TSS 11.0%～11.5%，酸1.5%，味浓偏酸，品质中上。种子10～12粒/果。成熟期11月下旬至12月上旬。该品种丰产性一般，较耐贮运。果皮与大红柑有同样的药用价值。

6. Hanggan

Origin and Distribution: Hanggan, or Yudonggan, Sihuigan, originated in Sihui, Guangdong and is grown a little in Zhaoqing. It has been introduced to Zhejiang, Hubei etc.

Main Characters: Tree semi-vigorous, spheroid, branches slender and dense. Leaf size 5.7~7.7cm×2.9~3.8cm, broad elliptic, margin shallowly dentate, with small petiole wing. Complete flower small. Fruit obloid, yellowish orange, size 6.6~7.6cm×5.1~5.3cm, apex truncate, slightly depressed, easily peeling. TSS 11.0%~11.5%, acid content 1.5%, rich but moderately sour flavor, Fruit quality is medium or superior. 10~12 seeds per fruits. Maturity period ranges from late November to early December. The variety is moderately productive, stores and ships comparatively well. The rind has same medicinal function as Dahonggan.

7. 红橘

来源与分布：红橘又名福橘、川橘、南橘、大红橘、江南橘，是我国古老品种。主产四川、重庆、福建，湖南、湖北、江西、浙江、云南、贵州等地有少量栽培。

主要性状：树势强健，树冠圆头形，半开张。幼树稍直立。枝梢细密，多刺。春梢叶片大小为6.8～7.3cm×2.8～3.2cm，椭圆形或近卵圆形，先端渐尖，凹口明显，基部楔形，翼叶线状。花小，单生或丛生。果实扁圆形，鲜红色，大小为6.0～6.5cm×4.5～4.8cm，油胞密，平生或微凸，果面光滑，果皮易剥离。TSS 11.0%～13.0%，酸0.6%～1.1%，品质中上。种子8～20粒/果。成熟期11～12月。该品种适应性强，丰产稳产，果实耐贮性差。砧木为枳或红橘。

7. Hongju (Red Tangerine)

Origin and Distribution: Hongju or Fuju, Chuanju, Nanju, Dahongju, Jiangnanju, is an old Chinese variety and currently grown in Sichuan, Chongqing and Fujian. There are small-scaled cultivation in Hunan, Hubei, Jiangxi, Zhejiang, Yunnan and Guizhou.

Main Characters: Tree vigorous, semi-spreading, spheroid, young trees somewhat erect. Branches slender and dense, thorny. Leaf 6.8~7.3cm×2.8~3.2cm, elliptic or nearly ovate, apex acuminate and notched, base cuniform, petiole wing linear. Flower small-sized, solitary or in cluster. Fruit obloid, bright red, sizes in 6.0~6.5cm×4.5~4.8cm, rind smooth; oil gland dense, flat, slightly rough; easily peeling. TSS 11.0%~13.0%, acid content 0.6%~1.1%, medium quality or superior. 8~20 seeds per fruit. Maturity period ranges from November to December. The variety is widely adaptable, a regular and productive bearer, stores not so well. The main rootstocks used are Trifoliate orange or Hongju.

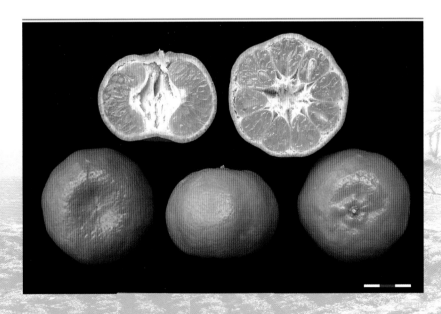

（1）**大红袍** 大红袍原产四川，重庆、湖北、贵州等地有栽培。

果实扁圆形，色泽鲜红，大小为6.0～6.2cm×4.5～4.7cm，果顶微凹，蒂部乳头凸起。TSS 11.5%～13.0%，酸0.4%～0.5%，品质中等。种子10粒/果。该品种适应性强，丰产稳产，果实不耐贮藏。

(1) Dahongpao　It is originated in Sichuan Province and grown in Chongqing, Hubei, Guizhou.

Fruit obloid, bright red, 6.0~6.2cm× 4.5~4.7cm, apex slightly depressed and base convex with a small nipple. TSS 11.5%~13.0%, acid content 0.4%~0.5%, of medium quality. 10 seeds per fruit. The variety is widely adaptable and productive.　Fruit does not store well.

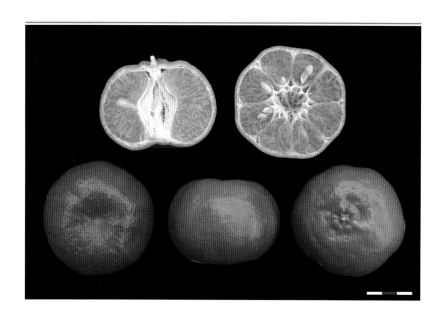

(2) **福橘** 福橘又名福州蜜橘，原产福建，主产福建，广东、广西、湖南、重庆等地有栽培。

果实扁圆形，鲜红色，果实大小为6.0～6.3cm×4.1～4.3cm，果顶微凹，果蒂部有凸起沟痕，果皮光滑，易剥离。TSS 11.0%～13.0%，酸0.8%～1.0%，果肉化渣，具微香。种子9～12粒/果。成熟期12月下旬。该品种适应性强，丰产稳产。已选出10月下旬至11月上旬成熟的早福橘。砧木用枳、红橘。

(2) Fuju Fuju or Fuzhoumiju originated in Fujian Province and is mainly grown there. Ther is limited cultivation in Guangdong, Guangxi, Hunan, and Chongqing.

Fruit obloid, bright red, 6.0~6.3cm×4.1~4.3cm, apex slightly depressed, base convex with grooves and ridges, rind smooth and easily peeling. TSS 11.0%~13.0%, acid content 0.8%~1.0%, pulp melting with weak aroma. 9~12 seeds per fruit. Matures in late December. The variety is widely adaptable, productive without alternate bearing. An early-maturing line has been selected which matures in late October to early November. The rootstocks used are Trifoliate orange and Hongju.

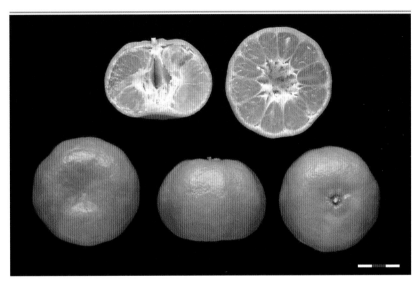

（3）**克里迈丁红橘** 克里迈丁红橘又名阿尔及利亚红橘。1965年我国从摩洛哥引入。在重庆、四川、广西、福建、湖北有少量栽培。

果实扁圆形，橙红色，有光泽，大小为5.2～5.5cm×3.4～3.9cm。TSS 10.0%～11.0%，酸0.7%～0.8%。种子14.4粒/果，单胚。品质中等。成熟期11月中旬。该品种单性结实力强，适应性好，丰产性一般，果实不耐贮藏，可作杂交育种材料。

(3) **Clementine** Clementine was introduced from Morocco in 1965 and is cultivated a little in Chongqing, Sichuan, Guangxi, Fujian and Hubei.

Fruit obloid, orange-red, glossy, 5.2~5.5cm×3.4~3.9cm, TSS 10.0%~11.0%, acid content 0.7%~0.8%. 14.4 seeds per fruit, monoembryonic. Quality medium. Matures in middle November. The variety is strongly parthenocarpic, very adaptable, moderately productive. Fruit does not store well. It can be used as mother parent in cross breeding.

(4) 南橘 南橘为中国的古老品种，主产湖南，广西等地有少量栽培。

树势中等，树冠较直立，春梢叶片大小为7.3～8.0cm×3.2～3.5cm，阔披针形。果实扁圆形，橙红色有光泽，大小为5.3～5.8cm×4.0～4.5cm。果皮易剥离。TSS 11.0%～13.0%，酸0.5%～0.9%，品质中等。种子20～30粒/果。成熟期10月下旬。该品种适应性强，丰产性一般，已选有少核品种。

(4) Nanju Nanju, an old Chinese variety, is mainly grown in Hunan and limitedly cultivated in Guangxi etc.

Tree medium-vigorous, relatively erect. Leaf 7.3~8.0cm×3.2~3.5cm, broad lanceolate. Fruit obloid, orange-red, glossy, 5.3~5.8cm×4.0~4.5cm, easily peeling. TSS 11.0%~13.0%, acid content 0.5%~0.9%, of medium quality. 20~30 seeds per fruit. Matures in late October. The variety is widely adaptable and moderately productive. Less-seeded selections are now available.

8. 建柑

来源与分布：建柑又名土橘、狗屎橘、药柑子。原产我国，长江流域各省（直辖市）有栽培和分布。

主要性状：树势中等，树冠圆头形，树姿开张。叶卵状椭圆形。花为单花。果实扁圆形，橙黄色，较粗糙，大小为6.2cm×4.6cm，果顶微凹，间有不明显印环，蒂部稍凸，果皮有特殊气味，易剥离。TSS 9.2%，酸0.5%，种子15～20粒/果。成熟期11月下旬至12月上旬。该品种适应性强，较丰产，主根发达，侧根分布均匀，可用作宽皮柑橘的砧木。

8. Jiangan

Origin and Distribution: Jiangan, or Tuju, Goushiju, Yaoganzi, originated in China and distributes in the provinces along Yangtze River.

Main Characters: Tree medium-vigorous, spheroid, spreading. Leaf ovate to elliptic. Solitary flower. Fruit obloid, 6.2cm×4.6cm, orange-yellow, surface moderately rough, apex slightly depressed, with inconspicuous areole ring, base somewhat convex, rind has distinctive odor, easily peeling. TSS 9.2%, acid content 0.5%, 15~20 seeds per fruit. Maturity period ranges from late November to early December. This variety has strong adaptability and is comparatively productive. Its strong primary root and evenly-distributed lateral roots make it a good rootstock for loose-skin mandarins.

9. 九月橘

来源与分布：九月橘原产广东四会，现四会有少量栽培，广西等地有引种。

主要性状：树冠圆头形，枝条细长。春梢叶片大小为5.9～6.9cm×2.2～3.9cm，椭圆形，叶缘锯齿，翼叶小。花小，完全花。果实扁圆形，橙红色，大小为5.0～5.7cm×3.5～4.0cm，果顶微凹，果蒂有浅放射沟纹，果皮易剥离。TSS 10.0%～11.0%，酸1.0%～1.2%，味稍酸，不太化渣，品质中等。种子10～15粒/果，成熟期11月上旬，该品种适应性强，丰产性中等。

9. Jiuyueju

Origin and Distribution: Jiuyueju originated in Sihui, Guangdong Province. There is limited cultivation in Sihui and some introduction in Guangxi etc.

Main Characters: Tree spheroid, branch long and slender. Leaf size 5.9~6.9cm× 2.2~3.9cm, elliptic, margin dentate, with small petiole wing. Complete flower, small-sized. Fruit obloid, orange-red, size 5.0~5.7cm×3.5~4.0cm, apex slightly depressed, base striped with radial furrows, easily peeling. Fruit quality is medium. TSS 10.0%~11.0%, acid content 1.0%~1.2%; taste slightly sour, not so melting; 10~15 seeds per fruit. Matures in Mid-November. The variety has strong adaptability and medium productivity.

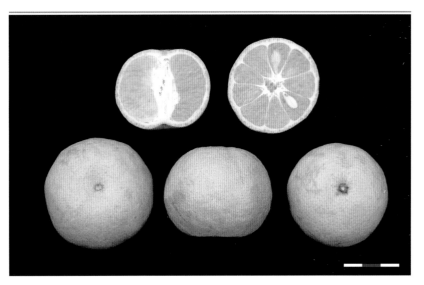

10. 马水橘

来源与分布：马水橘又名阳春甜橘，原产广东阳春，2003年通过广东省农作物品种审定委员会认定。主产广东云浮，广西等地有引种。

主要性状：树势健壮，树冠半圆头形，枝细密。春梢叶片大小为6.2～6.6cm×2.8～3.0cm，长椭圆形，秋梢叶缘锯齿明显，翼叶较小。花较小，完全花。果实扁圆形，橙黄色有光泽，大小为5.0～5.5cm×3.4～3.5cm，果顶平，微凹，果皮容易剥离。TSS 12.0%～13.0%，酸0.6%，味清甜较化渣。种子0～10粒/果，成熟期1月下旬至2月上旬。该品种丰产性好。主要砧木为酸橘、三湖红橘、枳。

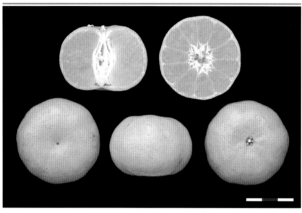

10. Mashuiju

Origin and Distribution: Mashuiju, or Yangchuntianju, originated from Yangchun, Guangdong Province. In 2003 it was registered as a new variety by Guangdong Crop Cultivar Registration Committee. There is large scale production in Yunfu, Guangdong and some introduction in Guangxi etc.

Main Characters: Tree vigorous, crown semi-spheroid, branches dense and slender. Leaf size 6.2~6.6cm×2.8~3.0cm, long-elliptic, leaves on autumn shoots dentate distinctively, petiole wing relatively small. Complete flower, comparatively small. Fruit obloid, surface glossy, size 5.0~5.5cm×3.4~3.5cm, apex truncate and slightly depressed, easily peeling. TSS 12.0%~13.0%, acid content 0.6%. Pulp pure sweet, comparatively melting. 0~10 seeds per fruit. Maturity period is from late January to early February. The variety is productive. The main rootstocks used are Sour tangerine, Sanhuhongju and Trifoliate orange.

11. 榣橘

来源与分布：榣橘原产浙江省黄岩，主产浙江黄岩等地。

主要性状：树势中等，树冠圆头形，枝条开张，无刺。春梢叶片大小为8.0～10.0cm×3.3～4.5cm，广椭圆形，叶面有高低不平皱纹，叶缘波状或有锯齿。花较小，为完全花。果实扁圆形，橙黄色，大小为5.0～6.0cm×4.2～4.5cm，蒂部突起，中心柱大而空，皮稍厚，易剥离。TSS 11.5%～12.0%，酸0.7%～0.8%，肉质细嫩，化渣。甜酸适中，品质中等。种子7～10粒/果。果实耐贮藏，经贮后味转甜。

11. Manju

Origin and Distribution: Manju originated in Huangyan, Zhejiang, and is mainly grown in Huangyan, Zhejiang Province.

Main Characters: Tree semi-vigorous, spheroid, branches spreading, thornless. Leaf size 8.0~10.0cm×3.3~4.5cm, broad elliptic, surface corrugated, margin sinuate or dentate. Complete flower, relatively small. Fruit obloid, yellowish orange, size 5.0~6.0cm×4.2~4.5cm, base protrude, axis large and hollow, rind slightly thick, easily peeling. TSS 11.5%~12.0%, acid content 0.7%~0.8%, pulp tender, melting with a moderate sweet sour taste, quality medium. 7~10 seeds per fruit. Fruit stores well and taste turns sweet after storage.

12. 明柳甜橘

来源与分布：明柳甜橘是广东省农业科学院果树研究所和紫金县科技局、杨坑柑橘场从春甜橘的芽变选出的新品系。2006年通过广东省农作物品种审定委员会审定。主产广东河源、惠州等市。

主要性状：树势强，树冠圆头形，枝条粗长有刺。春梢叶片大小为7.4～8.0cm×3.0～3.5cm，椭圆形，翼叶较小。花较小，完全花，自交不亲和。果实扁圆形，橙黄色，大小为5.4～5.8cm×4.0～4.5cm，果顶平，微凹，果面有柳纹，果皮容易剥离。TSS 12.0%～13.0%，酸0.4%～0.5%，无核。果肉汁多化渣，清甜有香味。成熟期比春橘迟6～7d。该品种容易保果，早结丰产性好，是我国最迟熟的甜橘品种。主要砧木为三湖红橘、酸橘等。

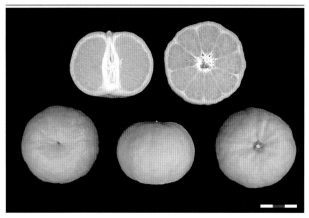

12. Mingliutianju

Origin and Distribution: Mingliutianju is a new bud line selected from Chuntianju, in Yangkeng Citrus Farm, by the Institute of Fruit Tree Research, Guangdong Academy of Agricultural Sciences and Zijin Sciences and Technology Bureau. In 2006 it was registered as a new variety by Guangdong Crop Cultivar Registration Committee. Its main production areas are Heyuan and Huizhou in Guangdong.

Main Characters: Tree vigorous with spheroid crown, branches long, hard and thorny. Leaf size 7.4~8.0cm×3.0~3.5cm, elliptic, petiole wing relatively small. Complete flower, comparatively small, self-incompatible. Fruit obloid, yellowish orange, size 5.4~5.8cm×4.0~4.5cm; apex truncate, slightly depressed; rind striped with willow leaf-like patterns, easily peeling. TSS 12.0%~12.7%, acid content 0.4%~0.5%, seedless. Pulp juicy, melting, pure sweet and fragrant. Its mature season is 6 to 7 days later than Chunju and is the latest among the sweet mandarins in China. The variety early fruiting, very productive and easy to prevent fruit drop. The main rootstocks used are Sanhuhongju and Sour tangerine.

13. 南丰蜜橘

来源与分布：南丰蜜橘又名金钱蜜橘、邵武蜜橘（福建）、茆橘，选自乳橘，是我国古老品种，有1300年栽培历史。原产江西省南丰，主产江西省南丰、临川等地，浙江、福建、湖南、湖北、四川、广西等省（自治区）有少量栽培。

主要性状：树势壮旺，树冠半圆头形，枝梢细长而稠密，无刺。春梢叶片大小为5.4～5.8cm×3.3～3.5cm，狭长卵圆形，叶缘锯齿较浅，翼叶较小。花较小，完全花。果实扁圆形，橙黄色，大小为3.3～3.5cm×2.5～3.5cm。果顶平，微凹，中心有小乳凸，果皮容易剥离。TSS11.0%～16.0%，酸0.8%～1.1%，种子0.7粒/果，一般1～2粒，成熟期11月上旬。该品种汁多，具浓郁香味，品质优，丰产性好，抗寒性强，易感疮痂病。主要砧木为枳。

通过品种整理和营养系选种，有大果系、小果系、桂花蒂系、早熟系等品系。

13. Nanfengmiju

Origin and Distribution: Nanfengmiju, or Jinqianmiju, Shaowumiju (Fujian), and Shiju, derived from Ruju. It is an ancient Chinese variety with a 1300 year-history of cultivation. It is native to Nanfeng, Jiangxi Province and is mainly grown in Nanfeng and Linchuan of Jiangxi Province. There are small-scale cultivations in Zhejiang, Fujian, Hunan, Hubei, Sichuan and Guangxi etc.

Main Characters: Tree quite vigorous, crown semi-spheroid, branches long, slender and dense, thornless. Leaf size 5.4~5.8cm×3.3~3.5cm, narrow ovate, margin shallow-dentate, petiole wing relatively small. Complete flower, comparatively small. Fruit obloid, yellowish orange, sizes in 3.3~3.5cm×2.5~3.5cm; apex truncate, slightly depressed, with a tiny nipple; easily peeling. TSS 11.0%~16.0%. acid 0.8%~1.1%. Seeds 0.7 per fruit, usually 1 or 2. Matures in early November. Abundant in juice, rich in aroma, and excellent in taste. The variety is quite productive, highly tolerant to cold, highly susceptive to citrus scab. The rootstock currently used is Trifoliate orange.

The variety contains lines of large-fruit, small-fruit, osmanthus-like calyx, and early-maturity by long-time clonal selection.

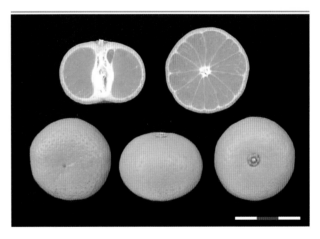

（1）**柳城蜜橘** 柳城蜜橘是从南丰蜜橘芽变中选育的新品种，1999年在广西柳州市柳城县凉水山林场发现。2012年通过广西农作物品种审定委员会审定。

树势强健，果实大，裂果少，丰产稳产。单果重38.4g，可食率为75.7%，TSS 13.2%，酸0.8%，味清甜，化渣，品质上等。10月下旬至11月下旬成熟。桂北地区及柳州都能栽培，适宜在南丰蜜橘种植的区域推广。

(1) Liuchengmiju Liuchengmiju is a new cultivar selected from a bud mutation of Nanfengmiju, which was found in Liangshuishan Forest Farm, Liucheng County, Liuzhou City, Guangxi in 1999, and registered in Guangxi in 2012.

Tree vigorous, fruit big and seldom cracking, prolific and yield stable. Fruit weight 38.4g, edible portion 75.7%. TSS 13.2%, TA 0.8%, flavor fresh sweet, quality first-class. Maturity from the end of October to the end of November. It is suitable for Northern Guangxi and Liuzhou, Guangxi; Production area for Nanfengmiju.

（2）**粤丰早橘** 粤丰早橘是南丰蜜橘的早熟优良变异，2004年在韶关市始兴县顿岗镇南丰蜜橘园发现，2014年通过广东省农作物品种审定委员会审定。

树势强，树姿较直立。结果母枝以春梢、秋梢为主。果实扁圆形，果面无沟纹，果顶稍凹，印圈较明显，果皮光滑，橙黄色，单果重58.3g，果皮厚0.18cm，种子0.3粒/果。TSS13.4%，总糖11.8%，酸0.5%，每100ml果汁维生素C含量20.8mg。果肉较化渣，品质优良。早结丰产性好，在广东地区果实10月上中旬成熟。适宜在粤北柑橘种植区推广。

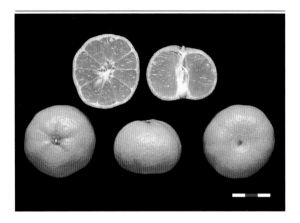

(2) Yuefengzaoju Yuefengzaoju is a bud mutation of Nanfeng miju, found in Shixing County, Shaoguan of Guangdong in 2004, and was registered as a new cultivar by Guangdong Crop Cultivar Registration Committee in 2014.

Tree vigorous, comparatively upright. Fruiting twigs are mainly spring or autumn shoots. Fruit oblate, surface without furrows, apex slightly depressed, with conspicuous areole ring, rind smooth, orange yellow in color. Fruit weight 58.3g, peel 0.18cm in thickness, nearly seedless with 0.3 seeds per fruit, TSS 13.4%, total sugar content 11.8%, TA 0.5%, and Vitamin C 20.8mg/100ml. Flesh has medium mastication. Quality excellent. Precocious and prolific. Maturity in early-to-mid October in Guangdong. It is suitable for the citrus cultivation area of northern Guangdong Province.

14. 年橘

来源与分布：年橘又名叶橘、潮汕橘、省橘，是我国古老品种。原产广东，原主产新会、中山，现主产广东龙门县。是广东近几年的主栽品种之一，广西等地有栽培。

主要性状：树势强，树冠圆头形，枝条细长而硬，有短刺。春梢叶片大小为6.5～7.2cm×3.2～3.8cm，狭长椭圆形，翼叶较小。花较小，完全花。果实扁圆形，橙黄色，大小为4.0～5.2cm×3.0～3.8cm，果顶平，微凹，果皮易剥离。TSS 10.0%～13.0%，酸0.8%～1.5%，种子12～15粒/果。成熟期1月中下旬，可树上挂果延迟至3月上旬。味偏酸。该品种丰产稳产性好，容易管理。主要砧木为酸橘等。

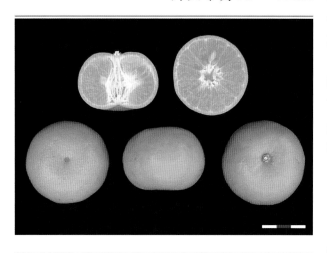

14. Nianju

Origin and Distribution: Nianju, or Yeju, Chaoshanju, Shengju, is an ancient variety originated in Guangdong. It was once widely grown in Xinhui and Zhongshan and is now mainly cultivated in Longmen of Guangdong. In recent years it has become one of the principal cultivars in Guangdong. There is some cultivation in Guangxi etc.

Main Characters: Tree vigorous, crown spheroid, branches long, slender and hard, with short thorns. Leaf size 6.5~7.2cm×3.2~3.8cm, long and narrow elliptic, petiole wing relatively small. Complete flower, comparatively small. Fruit obloid, yellowish orange, size 4.0~5.2cm×3.0~3.8cm, apex truncate, slightly depressed, easily peeling. TSS 10.0%~13.0%, acid content 0.8%~1.5%, 12~15 seeds per fruit. Matures in middle-late January. Fruit holds on tree well till early March. Fruit orange-red, somewhat sour. The variety is a regular and productive bearer without strict orchard management. The main rootstocks used are Sour tangerine etc.

少核年橘 少核年橘为年橘通过辐射诱导产生的少核品种，2014年通过广东省品种登记。

树势较壮旺，树冠圆头形，幼树较直立。果实扁圆形，果顶部平或稍凹，果面光滑，橙黄至深黄色，单果重51g，果皮厚0.17cm，易剥离，种子3.3粒/果，TSS14.5%，总糖10.3%，酸0.9%，每100ml果汁维生素C含量44.0mg，果肉汁胞柔软多汁、化渣，耐贮性较好。早结丰产，在广东地区果实成熟期为1月中下旬。适宜在广东柑橘种植区推广。

Shaohe Nianju Shaohe Nianju is a less-seedy mutant of Nianju induced by radiation, registered in Guangdong Province in 2014.

Tree vigorous, crown spheroid, comparatively upright in youth. Fruit obloid, apex truncate to slightly depressed, peel smooth, orange-yellow to dark yellow, average weight 51g, rind 0.17 cm thick, easy to peel, 3.3 seeds per fruit, TSS 14.5%, total sugar content 10.3%, TA 0.9%, and Vitamin C 44.0mg/100ml. Pulp tender, juicy and melting. Storage durable. Precocious and prolific. Mature in mid-to-late January in Guangdong. It is suitable for the citrus cultivation area in Guangdong Province.

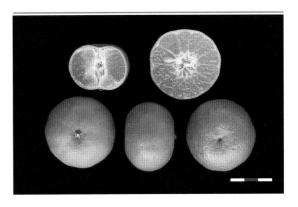

15. 椪柑

来源与分布：椪柑又名芦柑、有柑、蜜桶柑、潮州蜜橘、梅柑、白橘等，原产中国。主产福建、广东、台湾等地。广西、浙江、江西、湖南、四川、云南等地均有栽培，是中国分布最广的柑橘类型。

主要性状：树势强，树冠圆柱形或圆锥形，幼树枝条直立，老树稍开张，枝细而密集。春梢叶片大小4.5～5.5cm×2.0～2.5cm，卵圆形，叶缘呈波浪状，翼叶线状。花较小，完全花，多为单花。果实扁圆形或高扁圆形，果面橙黄色或橙红色，有光泽，果皮有松紧两种类型，易剥离，果实大小为7.0～8.5cm×6.0～7.5cm。TSS 11.0%～14.0%，酸0.5%～0.9%，种子5～10粒/果，果肉脆嫩，汁多，化渣，有香气，品质上等。成熟期11月上旬至翌年1月中旬。该品种丰产性较好，是我国宽皮柑橘主栽品种之一。主要砧木为枳、酸橘。

15. Ponkan

Origin and Distribution: Ponkan, also known as Lugan, Maogan, Mitonggan, Chaozhoumiju, Meigan and Baiju etc., originates from China and is one of the most widely distributed citrus variety in China. The main cultivating regions locate in the provinces of Fujian, Guangdong and Taiwan. Guangxi, Zhejiang, Jiangxi, Hunan, Sichuan and Yunnan provinces also produce Ponkan.

Main Characters: Tree vigorous, crown terete or conic shape; branches erect on young tree, slightly spreading on adult tree, slender and dense. Spring leaf size 4.5~5.5cm × 2.0~2.5cm, ovate, margin sinuate, petiole wing linear. Complete flower, relatively small, solitary. Fruit globose to moderately obloid; Rind color orange to deep-orange at maturity, loosely or moderately adherent and easily peeling; surface relatively smooth with prominent, sunken oil glands; Fruit size 7.0~8.5cm × 6.0~7.5cm. Fruit quality is excellent with TSS 11.0%~14.0%; acid 0.5%~0.9%; 5~10 seeds per fruit; flesh crisp, juicy, melting and fragrant. Matures from November to next January.

As one of the most widely-grown mandarins in China, Ponkan is comparatively productive. The main rootstocks for Ponkan are Trifoliate orange and Sour tangerine.

（1）东13椪柑　东13椪柑是广东省农业科学院果树研究所和杨村柑橘场选出的新品种。1989年通过广东省农作物品种审定委员会审定。

果实高扁圆形，橙红色，大小为8.1cm×6.5cm，果顶微凹，皮松易剥离。TSS 11.0%～13.0%，酸0.5%～0.7%，种子13粒/果，肉质脆嫩，高糖高酸，味浓，品质优。成熟期12月上旬。该品种早结丰产性好，果大美观、质优。

(1) Dong 13 Ponkan　Dong 13 Ponkan is a variety selected collaboratively by the Institute of Fruit Tree Research, Guangdong Academy of Agricultural Sciences and the Yangcun Citrus Farm, and was registered as a new cultivar by Guangdong Crop Cultivar Registration Committee in 1989.

Fruit long obloid, rind surface red-orange at maturity, size 8.1cm×6.5cm, apex slightly depressed. Rind loosely adherent, easily peeling. TSS 11.0%~13.0%, acid content 0.5%~0.7%, about 13 seeds per fruit. Flesh crisp and tender, high sugar and acid content, flavor rich with excellent quality. Matures in early December.

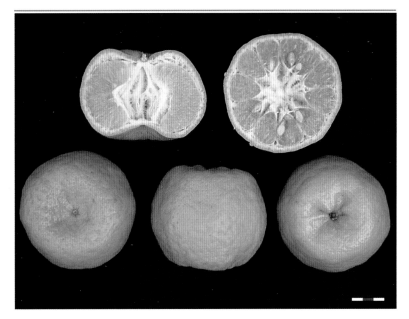

The variety has fairly well yield at the early fruiting stage and produces large, good-looking fruit with excellent quality.

(2) 鄂柑1号椪柑　鄂柑1号椪柑又名金水柑，是湖北省农业科学院果树茶叶研究所从实生苗中选出的新生系品种。

果实圆球形，橙黄色，大小为8.1cm×6.8cm，果顶微凹，果蒂突起有放射沟数条。TSS 12.0%，酸1.27%。成熟期11月中旬。该品种适应性强，耐低温，果实肉质脆嫩化渣，有香气。耐贮藏。

(2) E-gan No.1 Ponkan　E-gan No.1 ponkan, also named as Jinshuigan, was a nucellar line selected from seedlings by the Fruit and Tea Institute of Hubei Agricultural Science Academy.

Fruit spheroid, orange-yellow, size 8.1cm×6.8cm, apex slightly depressed, base slightly convex with several radial grooves. TSS 12.0%, acid content 1.27%. Matures in mid-November. The variety has wide adaptability and is tolerant to low temperature. Flesh crisp, melting and fragrant. Fruit has a good storage quality.

(3) **和阳2号椪柑** 和阳2号椪柑是原广东汕头柑橘研究所（现潮州市果树研究所）和福建省诏安县农业局在诏安选出。1988年通过广东省农作物品种审定委员会审定。

果实扁圆形，橙红色，大小为7.9cm×6.1cm，果顶凹入稍深。TSS 12.4%，酸0.9%，种子13.8粒/果。成熟期11月下旬至12月上旬。该品种丰产性好，果肉脆嫩化渣，汁多味浓、品质优。

(3) Heyang No.2 Ponkan It was selected from the orchard of Zhaoan County, jointly by former Guangdong Shantou Citrus Research Institute (Now is Chaozhou Fruit Institute) and the Agricultural Bureau of Zhaoan, Fujian Province. It was registered as a new variety by Guangdong Crop Cultivar Registration Committee in 1988.

Fruit obloid, red-orange, 7.9cm×6.1cm in size, with relatively deep depressed apex. TSS 12.4%, acid content 0.9%, 13.8 seeds per fruit. Maturity period ranges from late November to early December.

The variety is productive and its fruit has excellent quality with crisp, melting, juicy flesh of rich taste.

（4）**华柑2号椪柑** 华柑2号椪柑是华中农业大学、长阳土家族自治县农业技术推广中心等单位从老系实生硬芦园发现的优良单株，后经系统选育定名为清江椪柑1号，2005年通过湖北省农作物品种审定委员会审定，定名为华柑2号。

果实高扁圆形，橙黄至橙红色，单果重160g左右，果顶微凹，TSS 13.5%，酸0.7%，成熟期12月初。

(4) Huagan No.2 Ponkan An eminent line selected out from the old line of Yinglu seedling trees by Huazhong Agricultural University and the Agricultural Technology Extension Center of Changyang Tujia Autonomous County. It was originally named Qingjiang Ponkan No.1 after systematic selection and registered as a new variety under the name of Huagan No.2 by Hubei Crop Cultivar Registration Committee in 2005.

Fruit long obloid, yellowish orange to red-orange. Fruit average weights about 160g, apex slightly depressed. TSS 13.5%, acid content 0.7%. Matures in early December.

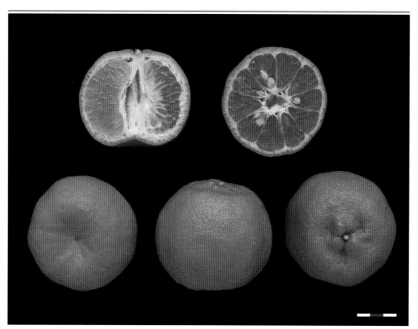

（5）**华柑4号椪柑** 华柑4号椪柑为长源椪柑的无籽芽变，1997年在江西省靖安县香田乡果园发现，2021年通过国家品种登记。

自然生长树形较直立，树势中强，萌芽力中等，一年抽梢3～4次。结果母枝以春梢和早秋梢为主，花芽分化能力强，花量中等，坐果能力中等。果实黄橙色，果面光滑，果皮较薄，易剥离，单果重125g。果肉脆嫩化渣。TSS 13.0%以上，酸0.9%，每100ml果汁维生素C含量30.0mg，可食率79.0%，混栽或单独栽培均无籽，为雌性败育类型，无籽性状十分稳定。在武汉地区果实完全成熟期为12月中下旬，较丰产。华柑4号与枳、资阳香橙、温州蜜柑及甜橙等嫁接亲和。最适栽培区为≥10℃的年有效积温超过5 800℃、极端最低气温≥－5℃的地区。

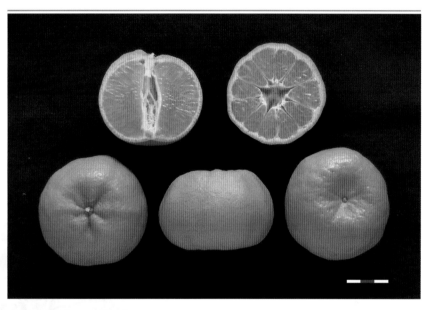

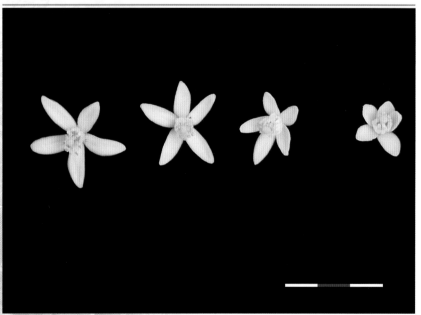

(5) Huagan No.4 Ponkan Huagan No.4 is a seedless bud mutation from the Changyuan Ponkan, found in Jing'an County of Jiangxi Province in 1997, and nationally registered by Huazhong Agricultural University in 2021.

Tree naturally upright, medium vigorous, with medium budding capacity of 3~4 times a year. Fruiting twigs are mainly spring or early autumn shoots. Strong flower bud differentiation capacity, flower quantity medium, fruit setting capacity medium. Rind smooth, yellow-orange, thin, easy to peel; average fruit weight 125g, pulp crisp, tender and melting, TSS 13.0%, TA 0.9% and Vitamin C 30.0mg/100ml, edible portion 79.0%. Constantly seedless in the single-cultivar or in mixed cultivation due to the female sterilization. In Wuhan area, fruits fully mature in mid-to-late December. Graft compatible with trifoliate orange, Ziyang junos, Satsuma and sweet orange. It is suitable for growing in areas with an annual effective accumulated temperature of $\geqslant 5\,800\,°C$ ($\geqslant 10\,°C$) and a minimum temperature $\geqslant -5\,°C$.

（6）黔阳无核椪柑 黔阳无核椪柑是湖南省洪江市科技局与湖南省园艺研究所从普通椪柑芽变中选出的新品系。1998年通过湖南省农作物品种审定委员会审定。

果实扁圆形，橙黄色，大小为7.1cm×6.5cm，果顶微凹，果蒂部有5～6条放射沟。TSS 13.5%，酸0.66%，无核。肉质脆嫩，汁多化渣。风味浓，品质优。该品种早结丰产，适应性广。

(6) Qianyang Seedless Ponkan It is a new bud line selected from common Ponkan by the Science and Technology Bureau of Hongjiang of Hunan Province and the Horticultural Research Institute of Hunan. Qianyang seedless ponkan was registered as a new variety by Hunan Crop Cultivar Registration Committee in 1998.

Fruit obloid, orange-yellow, size 7.1cm×6.5cm, apex slightly depressed, base convex, with 5~6 radial grooves. TSS 13.5%, acid content 0.66%, seedless. Fruit quality is excellent. Flesh crisp and tender, juicy and melting with rich flavor. The variety is early-fruiting, productive and has wide adaptability.

（7）试18椪柑　试18椪柑是广东省农业科学院果树研究所和杨村柑橘场选出的新品种。1990年通过广东省农作物品种审定委员会审定。

果实扁圆形，橙红色，大小为7.1～8.3cm×5.3～6.8cm，果顶平，微凹，果皮容易剥离。TSS 10.4%～11.3%，酸0.4%～0.6%，种子14粒/果。成熟期11月下旬至12月上旬。果肉柔软多汁，化渣，有香气，品质优。该品种丰产性好，早熟，果实大小较一致，商品率高。

（7）Shi 18 Ponkan　It was selected jointly by the Institute of Fruit Tree Research, Guangdong Academy of Agricultural Sciences and the Yangcun Citrus Farm, and registered as a new variety by Guangdong Crop Cultivar Registration Committee in 1990.

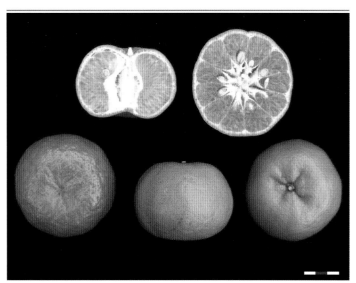

Fruit obloid, red-orange, size 7.1~8.3cm×5.3~6.8cm, apex truncate to slightly depressed, easily peeling. TSS 11.0%~13.0%, acid content 0.5%~0.7%, 14 seeds per fruit. Matures from late November to early December. Flesh tender, juicy, melting, and fragrant, of fine quality. The variety is productive and early maturing. Fruit size is relatively uniform, resulting in a higher commodity ratio.

(8) **新生系3号椪柑** 新生系3号椪柑源自四川江津园艺试验场引入的广东潮汕椪柑种子播出的实生苗。其中被中国农业科学院柑橘研究所收集在国家柑橘种质资源圃中的一株，经多年观察表现树体健康，品质优良，后定名为新生系3号椪柑。

果实高扁圆形，橙黄色，大小为6.0～7.2cm×5.4～6.8cm，果顶微凹，果蒂部稍有几条放射沟。TSS 11.5%～12.5%，酸0.6%。种子7～9粒/果。成熟期12月上中旬。耐贮藏。

(8) **Xinshengxi No.3 Ponkan** One of the seedlings, propagated from seeds introduced from Chaoshan, Guangdong Province by Jiangjin Horticultural Experiment Farm of Sichuan Province, was by chance collected by the Citrus Research Institute, Chinese Academy of Agricultural Sciences into the National Citrus Germplasm Repository and proven that the tree was healthy with very good fruit quality after many years'observation, and was later given the current name.

Fruit long obloid, orange-yellow, size 6.0~7.2cm×5.4~6.8cm, apex slightly depressed, base slightly convex with several radial grooves. TSS 11.5%~12.5%, acid content 0.6%,7~9 seeds per fruit. Matures in early-middle December. Fruit has a good storage quality.

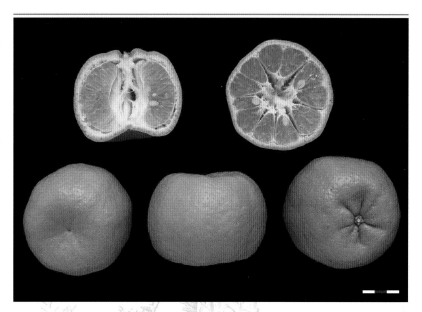

（9）**岩溪晚芦椪柑** 岩溪晚芦椪柑是福建长泰县农业局和该县岩溪青年果林场从普通芦柑变异单株选育而成的品种。已通过全国新品种审定。

果实扁圆形，橙黄色，大小为7.5cm×5.5cm，果顶较平至微凹，有较明显的放射沟8～11条。TSS 12.0%～15.0%，酸0.8%～1.2%，种子5.6粒/果。成熟期1月下旬至2月上旬。该品种品质优良，种子少，耐贮藏，少裂果，抗寒性好。

(9) Yanxiwanlu Ponkan Yanxiwanlu ponkan is a chance mutation selected from common Lugan by the Agricultural Bureau of Changtai County, Fujian Province and the local Yanxi Youth Fruit Tree Farm. It has passed the national new cultivar registration in China.

Fruit obloid, orange-yellow, size 7.5cm×5.5cm, apex truncate to slightly depressed, with 8~11 moderately prominent radial grooves. TSS 12.0%~15.0%, acid content 0.8%~1.2%, 5.6 seeds per fruit. Matures from late January to early February. The variety is of fine quality with less seeds, strongly tolerant to storage, less fruit splitting, and cold tolerant.

（10）**永春椪柑** 永春椪柑是福建省永春县20世纪50年代从福建漳州市引入，经长期栽培选育而成的品种。目前永春县是我国最大的椪柑生产基地。

果实扁圆形，橙红色，大小为5.5cm×7.1cm，果顶微凹入。TSS12.0%～15.0%，酸0.8%～1.0%。种子5～10粒/果。肉质脆嫩、化渣、多汁，甜酸适中。成熟期11月下旬至12月上旬。该品种早结，丰产稳产，品质优。

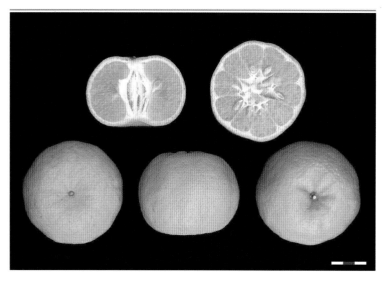

(10) Yongchun Ponkan Yongchun ponkan was selected in Yongchun County from the original cultivar introduced from Zhangzhou of Fujian Province at 1950s. Today Yongchun County is the largest Ponkan production base in China.

Fruit obloid, red-orange, 5.5cm× 7.1 cm, apex slightly depressed. TSS 12.0%~15.0%, acid content 0.8%~1.0%, 5~10 seeds per fruit. Flesh crisp, melting and juicy, moderate sweet sour. Matures from late November to early December. The variety is early fruiting, productive, less prone to alternate bearing, and excellent in fruit quality.

椪柑生产基地　Ponkan production base

(11) **粤优椪柑** 粤优椪柑（原为85-1椪柑）是广东省农业科学院果树研究所等单位从台湾省椪柑引种材料中通过系统选育选出。2006年通过广东省农作物品种审定委员会审定。

果实高扁圆形，大小为8.4cm×7.2cm，果顶平，微凹，皮松易剥离。TSS 12.2%，酸0.9%，种子4.5粒/果。该品种丰产性好，汁多化渣，味浓，品质优。

(11) Yueyou Ponkan Previously named as 85-1 Ponkan, Yueyou ponkan was systematically selected from Taiwan ponkan introduced before by the Institute of Fruit Tree Research, Guangdong Academy of Agricultural Sciences. It was registered as a new variety by Guangdong Crop Cultivar Registration Committee in 2006.

Fruit long obloid, size 8.4cm×7.2cm, apex truncate to slightly depressed. Easily peeling. TSS 12.2%, acid content 0.9%, 4.5 seeds per fruit. This variety is productive and has excellent fruit quality of juicy, melting and rich taste.

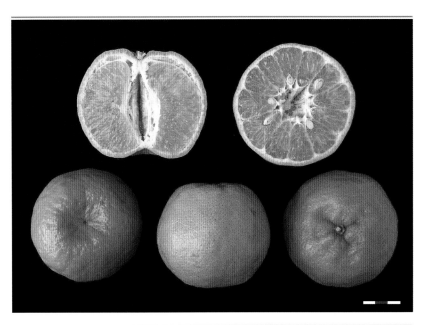

（12）**早蜜椪柑**　早蜜椪柑为辛女椪柑芽变，2004年在湘西州泸溪县良家潭乡家岩寨村果园发现，2012年通过湖南省农作物品种审定委员会审定。

树势强健，树形较直立，成枝力强，枝条密，嫩梢光滑、淡绿色，枝条无刺。单花，无花序，结果母枝以春梢或早秋梢为主。果实扁圆形，果形指数0.76，果沟明显，果顶部分有脐。果面光滑、果皮橘红，皮薄易剥离，厚度0.16～0.24mm。单果重124g，果肉细嫩化渣，风味浓，TSS13.2%以上，酸0.7%，每100ml果汁维生素C含量为47.1mg，可食率76.1%。自然授粉条件下种子3粒/果，单品种栽培表现少籽或无籽。早蜜椪柑与枳、温州蜜柑及其他橙类均嫁接亲和，在湘西地区成熟期11月上旬，比辛女椪柑早20～30d。最适栽培区为中亚热带柑橘种植区。

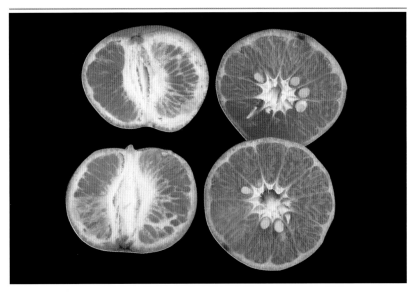

(12) Zaomi Ponkan　Zaomi Ponkan, a bud mutation of Xinnü Ponkan, was found in 2004 in the orchard of Luxi County, Xiangxi Prefecture of Hunan, and registered in Hunan in 2012.

Tree vigorous, somewhat erect with strong and dense branches. The young shoots smooth, light green and thornless. Solitary flowers, no inflorescence; mainly spring or early autumn shoots bear fruits. Obloid fruit with a shape index of 0.76, conspicuous fallows and apex with a navel. Orange-red skin, smooth and easy to peel thin with a thickness of 0.16~0.24cm. Average fruit weight 124g, pulp tender and melting with rich flavor, TSS above 13.2%, TA 0.7%, Vitamin C is 47.1mg/100ml, edible portion 76.1%. Natural pollination results in about three seeds per fruit, while single-variety cultivated fruit bears few seeds or no seed. Zaomi Ponkan is graft-compatible with Trifoliate Orange, Satsuma Mandarin and other sweet oranges. Zaomi matures in early November in western Hunan, 20~30 days earlier than Xinnü Ponkan. Most suitable in central subtropical growing areas for citrus.

16. 椪橘

来源与分布：椪橘原产广东潮汕，现广东河源有栽培。

主要性状：树势中等，春梢叶片大小为6.2cm×4.4cm，椭圆形。果实高扁圆形，橙红色，大小为4.5～5.0cm×3.6～3.8cm，果顶平，有印圈，果蒂部有放射纹，果皮易剥离。TSS 11.0%～12.0%，酸0.9%，酸甜，品质中等，种子8～10粒/果。成熟期12月中下旬，可挂果至春节前后上市。该品种早结丰产性好，砧木为酸橘。

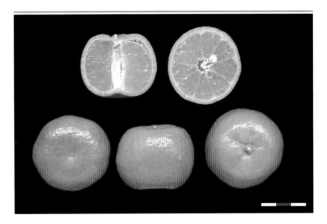

16. Pengju

Origin and Distribution: Pengju, originated in Chaoshan, Guangdong, is currently cultivated in Heyuan, Guangdong.

Main Characters: Tree medium-vigorous. Leaf 6.2cm×4.4cm in size, elliptic. Fruit long obloid, orange-red, size 4.5~5.0cm×3.6~3.8cm, apex truncate with areole ring, base striped with radial grooves, easily peeling. TSS 11.0%~12.0%, acid content 0.9%, sweet and acid flavor, medium quality. 8~10 seeds per fruit. Matures between middle and late December. Fruit can hold on tree till marketing during Spring Festival. The variety is early fruiting, very productive. Its suitable rootstock is Sour tangerine.

17. 乳橘

来源与分布：我国古老品种，在唐代已作为贡品，《新唐书》（1060年）中载有台州土贡乳柑。宋代《橘录》中的真柑（即乳橘），当时被韩彦直评为最优良的品种。产于浙江黄岩、温州，江西南丰、临川一带栽培的南丰蜜橘均属之。福建、江西、湖南、湖北、四川等省有少量栽培。

主要性状：树较矮，半圆头形，枝梢细密，无刺。春梢叶片大小为6.4cm×2.5cm，小卵状椭圆形。果小，扁圆形，橙黄色，大小为4.7～5.8cm×3.0～4.2cm。果皮薄，易剥离。TSS 12.0%～14.0%，酸0.3%～0.4%，果肉香甜多汁，柔软化渣，品质极佳。成熟期11月上中旬。该品种丰产，品质优。不耐贮藏。

17. Ruju

Origin and Distribution: Ruju is an ancient cultivar and was listed as tribute in the Tang Dynasty. There was record in *New Book of Tang* (1060A.D.) that the tribute Ruju was grown in Taizhou. Zhengan (Ruju) was considered as the best cultivar in the Song Dynasty by Han Yanzhi in the book *Ju Lu*. All "Nanfengmiju"mandarins, grown in Huangyan and Wenzhou of Zhejiang Province, and Nanfeng and Linchuan of Jiangxi Province, belong to this variety. Fujian, Jiangxi, Hunan, Hubei and Sichuan and other provinces have small-scale cultivations.

Main Characters: Tree comparatively small, crown semi-spheroid, branches slender and dense, thornless. Leaf size 6.4cm×2.5cm, small ovate-like elliptic. Fruit small-sized, 4.7~5.8cm×3.0~4.2cm, obloid, yellowish orange. Rind thin and easily peeling. TSS 12.0%~14.0%, acid content 0.3%~0.4%, pulp sweet, fragrant, juicy, tender and melting, taste excellent. It matures in early-middle November. The variety is productive and excellent in quality, though does not store well.

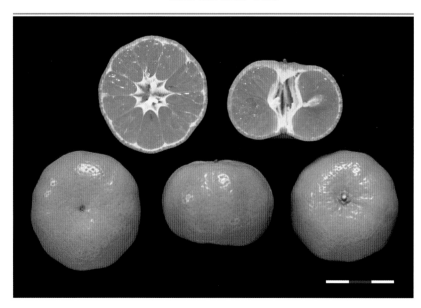

18. 砂糖橘

来源与分布：砂糖橘在20世纪90年代前称十月橘、冰糖橘，原产广东四会。是我国20世纪末发展的宽皮柑橘主栽品种之一，主产广东，广西、福建、江西、四川有分布。华南农业大学选育的无核砂糖橘，2006年通过广东省农作物品种审定委员会审定。

主要性状：树势健壮，树冠圆头形，枝条细密，稍直立。春梢叶片大小为7.5～8.2cm×4.3～5.2cm，椭圆形，叶缘锯齿稍深，翼叶较小。花较小，完全花。果实扁圆形，果皮油胞粗而突出，橙黄至橙红色，大小为4.5～5.5cm×3.5～4.2cm，果顶平而微凹，果皮容易剥离。TSS 12.0%～14.0%，酸0.3%～0.5%，种子0～12粒/果。成熟期11月下旬至12月上旬。该品种早结丰产性好，果肉细嫩，汁多味浓甜，品质上等。贮藏性稍差。主要砧木为酸橘、三湖红橘、枳。

18. Shatangju

Origin and Distribution: Shatangju, named as Shiyueju, Bingtangju before 1990s, originated in Sihui of Guangdong Province. It is one of the most newly grown varieties during the last decade in China, and mainly planted in Guangdong. Guangxi, Fujian, Jiangxi and Sichuan also has its distribution. The seedless line was selected by Huanan Agricultural University and registered as a new cultivar by the Guangdong Cultivar Registration Committee in 2006.

Main Characters: Tree vigorous with spheroid crown, branches slender and dense, somewhat erect. Leaf size 7.5~8.2cm×4.3~5.2cm, elliptic, with slightly deep dentate margin, wing relatively small. Complete flower, comparatively small. Fruit obloid, oil gland conspicuous and coarse, orange-yellow to orange-red, sizes in 4.5~5.5cm×3.5~4.2cm, apex truncate and slightly depressed, easily peeling. TSS 12.0%~14.0%, acid content 0.3%~0.5%, 0~12 seeds per fruit. Maturity period from late November to early December. The variety is early in maturity, very productive. Fruit has superior quality with flesh of tender and juicy and rich sweet. Storage performance slightly disappointing. Commonly used rootstocks are Sour tangerine, Sanhuhongju, Trifoliate orange.

中国柑橘品种 Citrus Varieties in China

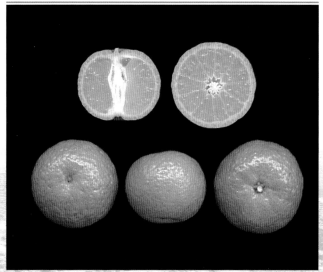

无核砂糖橘 Seedless Shatangju

无核砂糖橘 Seedless Shatangju

19. 酸橘

（1）**红皮酸橘**　红皮酸橘原产我国，有海丰红皮酸橘等。广西、广东、湖南等省（自治区）均有栽培。

树势较强，枝条较粗，嫩梢紫红色，树冠圆头形。春梢叶片大小为6.8～7.0cm×3.0～3.2cm，长椭圆形。花小，完全花。果实扁圆形，橙红色，大小为3～3.4cm×2.5～2.7cm，果顶凹入稍深，果蒂部有5～6条放射沟纹，果皮易剥离。TSS 10.0%～10.5%，酸2.2%～2.5%，种子14～15粒/果。成熟期12月上中旬。是蕉柑、椪柑、甜橙的良好砧木，嫁接后丰产性好，但结果稍晚。

19. Suanju (Sour Tangerine)

(1) Hongpisuanju　Originated in China, the represented cultivar is Haifeng Hongpisuanju. Its cultivated areas mainly distribute in Guangxi, Guangdong and Hunan etc.

Tree comparatively vigorous, spheroid, branches relatively thick, young twigs tinged with purple-red. Leaf 6.8~7.0cm×3.0~3.2cm, long-elliptic. Complete flower, small. Fruit obloid, orange-red, 3~3.4cm×2.5~2.7cm, apex moderately depressed, base striped with 5～6 radial furrows, easily peeling. TSS 10.0%~10.5%, acid content 2.2%~2.5%, 14~15 seeds per fruit. Matures in early December. It is a good rootstock of Jiaogan, Ponkan and Sweet oranges, and makes these scion cultivars very productive though slightly late to bear fruit.

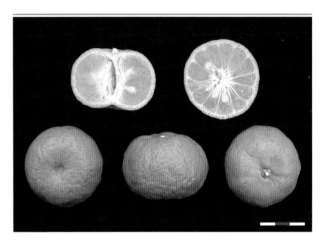

（2）**软枝酸橘** 软枝酸橘原产我国广东潮汕，广东、广西等地有栽培。

长势中等，树冠圆头形，枝条细软密生。春梢叶片大小为7.0～7.5cm×2.2～2.6cm，狭长椭圆形。花小，完全花。果实扁圆形，橙黄色，大小为3.8～4.2cm×2.8～3.2cm，果顶凹入稍深，果皮易剥离。TSS 10.0%～10.5%，酸2.0%～2.5%，种子15～17粒/果。成熟期12月上旬。是甜橙、蕉柑、椪柑等的良好砧木。嫁接后早结丰产性好。

(2) Ruanzhisuanju Ruanzhisuanju is native to Chaoshan, Guangdong, China. Its cultivated areas mainly distributes in Guangdong, Guangxi etc.

Tree medium-vigorous, spheroid, branches slender, soft and dense. Leaf 7.0~7.5cm×2.2~2.6cm, narrow, long-elliptic. Complete flower, small. Fruit obloid, orange-yellow, 3.8~4.2cm×2.8~3.2cm, apex moderately depressed, easily peeling. TSS 10.0%~10.5%, acid content 2.0%~2.5%, 15~17 seeds per fruit. Matures in early December. It is a good rootstock of Sweet orange, Jiaogan and Ponkan, and makes the scion varieties early-fruiting and productive.

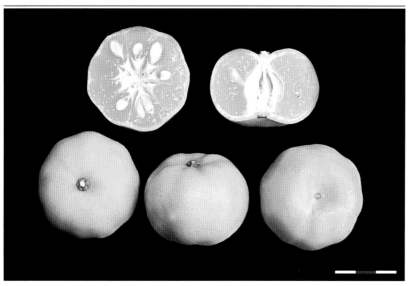

20. 韦尔金橘

来源与分布：韦尔金橘20世纪60年代从摩洛哥引入我国，现广东中山、惠州等市有栽培。

主要性状：树势中等，树冠圆头形或圆柱形，枝细而密，具有短小刺。春梢叶片大小为8.2～9.5cm×3.2～3.7cm，披针形。花中等大小，完全花。果实高扁圆形，橙黄色，有光泽，大小为5.5～6.4cm×4.4～5.2cm，果顶平，微凹，果皮易剥离。TSS 11.0%～13.0%，酸0.7%，果实肉质细嫩化渣，品质优。种子12粒/果。该品种丰产性好，主要砧木为酸橘。

20. Wilking

Origin and Distribution: Wilking was introduced to China from Morocco in the 1960s and is currently grown in Zhongshan and Huizhou of Guangdong Province.

Main Characters: Tree medium-vigorous, crown spheroid or columniform, branches slender and dense, with short thorns. Leaf 8.2~9.5cm×3.2~3.7cm, lanceolate. Complete flower, medium-sized. Fruit long obloid, orange-yellow, glossy, 5.5~6.4cm×4.4~5.2cm in size; apex truncate slightly depressed; easily peeling. TSS 11.0%~13.0%, acid content 0.7%; pulp tender and melting, of excellent quality. 12 seeds per fruit. The variety is quite productive. The main rootstock used is Sour tangerine.

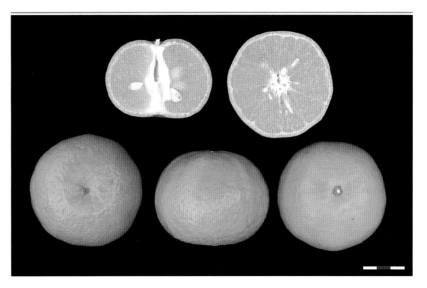

21. 五月橘

来源与分布：五月橘又名蜜橘，原产广东四会，四会等地有少量栽培。

主要性状：树势中等，春梢叶片大小为5.5～6.2cm×2.4～4.9cm，椭圆形。果实扁圆形，橙黄色，大小为5.1～5.5cm×4.1～4.6cm，果顶平，微凹，果皮容易剥离。TSS 11.0%～12.0%，酸0.6%，种子6～9粒/果。农历5月其果肉微甜可食，故名"五月橘"。成熟期11月中下旬。主要砧木为酸橘等。

21. Wuyueju

Origin and Distribution: Wuyueju, or Miju, originated in Sihui, Guangdong Province and grows a little in Sihui.

Main Characters: Tree semi-vigorous, leaf size 5.5~6.2cm×2.4~4.9cm, elliptic. Fruit obloid, orange-yellow, size 5.1~5.5cm×4.1~4.6cm, apex truncate, slightly depressed, easily peeling. TSS 11.0%~12.0%, acid content 0.6%, 6~9 seeds per fruit. In May of Chinese lunar calendar, pulp becomes slightly sweet and edible, and hence named as "Wuyueju" (Tangerine of May). Matures in middle to late November. The main rootstock is Sour tangerine.

22. 阳山橘

来源与分布：阳山橘原产广东阳山，主产阳山、新会、惠州、清远等市。

主要性状：树势较强，树冠圆头形，枝细长。春梢叶片大小为7.0～7.5cm×3.0～3.4cm，阔披针形或长椭圆形，叶缘锯齿不明显，翼叶较小。花较小，完全花。果实扁圆形，橙黄色，大小为6.0～6.4cm×4.5～4.8cm。TSS 11.0%～13.0%，酸0.7%～0.9%，种子16～25粒/果。果肉细嫩，汁多。成熟期1月中下旬。该品种早结丰产性好。主要砧木为酸橘、三湖红橘等。

22. Yangshanju

Origin and Distribution: Yangshanju, originated in Yangshan, Guangdong; is mainly grown in Yangshan, Xinhui, Huizhou and Qingyuan etc.

Main Characters: Tree comparatively vigorous, crown spheroid, branch long and slender. Leaf size 7.0~7.5cm×3.0~3.4cm, broad lanceolate or long elliptic, margin faintly dentate, petiole wing relatively small. Complete flower, comparatively small. Fruit obloid, orange-yellow, size 6.0~6.4cm× 4.5~4.8cm. TSS 11.0%~13.0%, acid content 0.7%~0.9%, 16~25 seeds per fruit. Pulp tender and juicy. Matures from middle to late January. The cultivar is early fruiting and very productive. The main rootstocks used are Sour tangerine and Sanhuhongju.

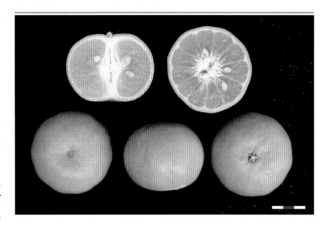

23. 早橘

来源与分布：早橘又名黄岩蜜橘，原产浙江黄岩。主产浙江台州。

主要性状：树势强健，树冠不规则半圆形，枝梢较疏朗，大枝直立，小枝少。春梢叶片大小为6.5～7.0cm×2.5～3.0cm，长椭圆形，先端微尖，基部宽楔形。花稍大，完全花。果实扁圆形，橙黄色，大小为5.4～5.8cm×3.6～4.0cm，油胞极小而密生，果皮易剥离。TSS 10.0%～11.0%，酸0.8%～0.9%，种子6～8粒/果，味酸甜，品质不如本地早，不耐贮藏。成熟期10月中下旬，果皮半黄即可采收。该品种产量高，早熟，既可鲜食也可加工成糖水橘瓣罐头，砧木是枸头橙。

23. Zaoju

Origin and Distribution: Zaoju, or Huangyanmiju, originated in Huangyan, Zhejiang Province, is mainly grown in Taizhou, Zhejiang.

Main Characters: Tree vigorous, with irregular semi-spheroid crown, branches somewhat sparse and spreading, main branches erect, twigs sparse. Leaf size 6.5~7.0cm×2.5~3.0cm, long-elliptic, apex slightly acuminate, base broad cuniform. Complete flower, slightly large. Fruit obloid, orange-yellow, size 5.4~5.8cm×3.6~4.0cm, oil glands extremely small and dense. Easily peeling. TSS 10.0%~11.0%, acid content 0.8%~0.9%, 6~8 seeds per fruit. Taste sour sweet. Quality is not as good as Bendizao. Fruit matures in middle-late October and can be harvested when half-yellow. It does not store well. The variety is productive and early-maturing, and good for fresh and canned consumption. Its rootstock is Goutoucheng.

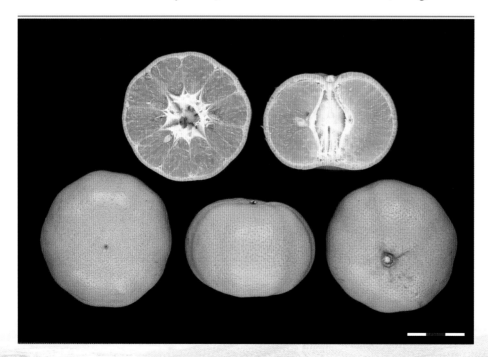

24. 朱橘

（1）**满头红**　满头红原产浙江，是朱红橘的实生变异，浙江、江西、广西有少量栽培。

树冠自然圆头形，枝疏而粗长。叶小，狭椭圆形。花稍大，完全花。果实扁圆形，朱红色，较光滑，大小为5.6～5.8cm×4.6～4.8cm。TSS 10.5%，酸0.7%～0.8%，种子8粒/果。成熟期11月下旬。该品种适应性强，丰产性较好，品质优。

24. Zhuju

(1) Mantouhong   Originated in Zhejiang, it was a chance seedling of Zhuhongju, and is grown in a small scale in Zhejiang, Jiangxi and Guangxi.

Tree natural spheroid; branches sparse, sturdy and long; Leaf small, narrow elliptic. Complete flower, somewhat large. Fruit obloid, rind deep red, relatively smooth, size 5.6~5.8cm×4.6~4.8cm, TSS 10.5%, acid content 0.7%~0.8%, 8 seeds per fruit. The variety is quite adaptable and comparatively productive, of excellent quality.

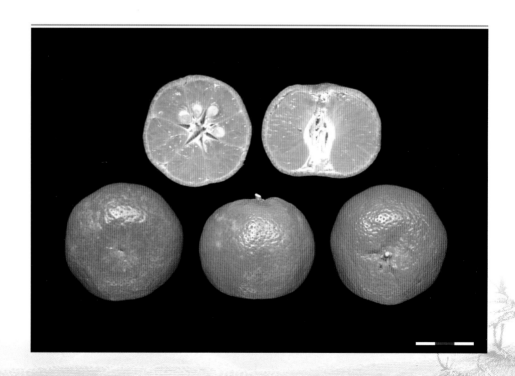

（2）**牛肉红朱橘** 牛肉红朱橘是从贵州省惠水县实生金钱橘（朱橘）中选育的新品种。2011年通过贵州省农作物品种审定委员会审定。

树势中庸，树冠开张，不规则圆头形，叶片较小而浓绿。花小，乳白色。果实圆球形至扁圆形，单果重52.7g，果皮深红色，果肉红色，囊瓣7～10瓣，单一栽培时种子2.4粒/果。TSS13.8%，每100ml果汁含糖10.9g、酸1.0g、维生素C 31.5mg。11月下旬成熟。该品种适应性广，在我国柑橘主产区都可栽培，但成熟季节冷凉气候有利于着色。

(2) Niurouhong Zhuju Niurouhong Zhuju is a new selection from chance seedling variation found in Huishui County of Guizhou, registered in this province in 2011.

Tree medium in vigor, crown spreading, irregular spheroid. Leaf small, dark green; flower small, white. Fruit spheroid to obloid, weight 52.7g, rind dark red, pulp red, segments 7~10, TSS 13.8%, TA 1.0%, Vitamin C 31.5mg/100ml. When grown as a single cultivar, 2.4 seeds per fruit. Mature in late November. This widely adaptable variety can be grown in most of the citrus production areas. Cool-weather in the maturing season favors color formation.

（3）**三湖红橘** 三湖红橘已有800多年的栽培历史，有九月黄和八月黄等。原产江西新干，江西新干、清江（现樟树市）等地有少量栽培。

九月黄长势中等，树冠圆头形，枝条细长。春梢叶片大小为8.3～8.7cm×3.3～3.8cm，长椭圆形。果实扁圆形，朱红色，大小为4.5～4.8cm×3.5～3.7cm，果顶平，微凹，果皮易剥离。TSS 10.0%～11.5%，酸0.5%～0.6%。种子9～10粒/果，农历九月成熟，品质优，丰产。八月黄农历八月成熟，品质稍次，种子较多，丰产。是蕉柑、椪柑、砂糖橘、甜橙的良好砧木，嫁接后早结、丰产、稳产，品质好，且耐裂皮病、碎叶病。

(3) Sanhuhongju With over 800-year history of cultivation, Sanhuhongju, originated in Xingan, Jiangxi, can be divided into different groups such as Bayuehuang, Jiuyuehuang etc. Now, Xingan and Zhangshu of Jiangxi Province have small-scale cultivation of Sanhuhongju.

Tree medium-vigorous, spheroid, branches long and slender. Leaf 8.3~8.7cm×3.3~3.8cm, long-elliptic. Fruit obloid, deep red, 4.5~4.8cm×3.5~3.7cm; apex truncate, slightly depressed; easily peeling. TSS 10.0%~11.5%, acid content 0.5%~0.6%, 9~10 seeds per fruit. Matures in lunar September. The variety of Jiuyuehuang is productive with superior fruit. Bayuehuang matures in lunar August, quality slightly lower than Jiuyuehuang, seedy and productive. Sanhuhongju mandrains are good rootstocks for Jiaogan, Ponkan, Shatangju and Sweet orange. Trees on Sanhuhongju are early-fruiting, productive, bear fruit of superior quality, and are tolerant to citrus exocortis viroid (CEVd) and citrus tatter leaf viroid (CTLV).

八月黄 Bayuehuang

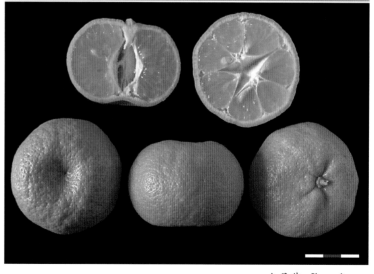

九月黄 Jiuyuehuang

（4）**朱红橘** 朱红橘原产我国，是我国古老品种，主产江苏、江西、陕西、湖北、湖南、四川、重庆等地，安徽、甘肃、河南有少量栽培。

树势强健，树冠不规则圆头形，大枝粗长稀疏，微向下披垂，小枝细密。幼树稍直立。春梢叶片大小为6.0～6.5cm×2.8～3.2cm，先端渐尖，基部广楔形，叶缘波状有不明显的锯齿。花小，完全花。果实扁圆形，朱红色，大果的果顶微凹，小果的果顶有乳头状凸起。表皮稍粗，果皮易剥离。TSS 10.0%～12.0%，酸0.8%～0.9%。种子较多，6～15粒/果。成熟期11月下旬至12月上旬。该品种易栽易管，丰产稳产，抗寒性、适应性强，品质中等。已选育有少核和无核朱红橘。

(4) **Zhuhongju** Zhuhongju, an old Chinese citrus variety, is mainly grown in Jiangsu, Jiangxi, Shaanxi, Hubei, Hunan, Sichuan and Chongqing, and limitedly cultivated in Anhui, Gansu and Henan.

Tree vigorous, irregular spheroid, main branches sturdy and long, sparse, slightly drooping; branchlets slender and dense. Young trees relatively erect. Leaf 6.0~6.5cm×2.8~3.2cm, apex acuminate, base broad cuniform, margin sinuate, faintly dentate. Complete flower, small. Fruit obloid, rind deep red. Apex of larger fruit slightly depressed, and of smaller fruit with a nipple-like structure. Rind slightly rough, easily peeling. TSS 10.0%~12.0%, acid content 0.8%~0.9%. Seedy, 6~15 seeds per fruit, quality medium. Maturity period ranges from late November to early December. The variety is easy to plant, productive, cold hardy, and very adaptable. Less-seedy and seedless lines have been selected.

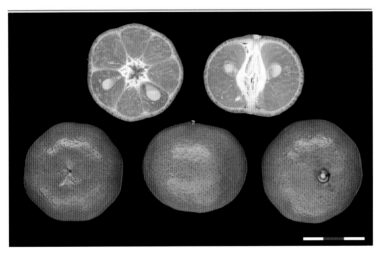

朱红橘　Zhuhongju

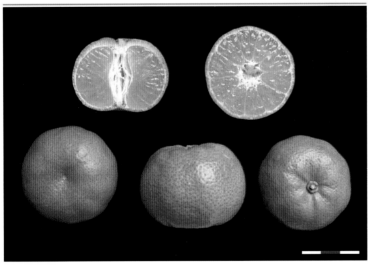

无核朱红橘　Seedless Zhuhongju

（5）**朱砂橘** 朱砂橘原产广东，主产广东，广西等地有少量栽培。

树冠圆头形，枝条细密。春梢叶片大小为7.2～8.0cm×3.8～4.2cm，卵圆形。花小，完全花。果实扁圆形，朱红色有光泽，果顶及果蒂部微凹，果皮易剥离。种子15～20粒/果，味酸甜。盆栽果大小为6.5cm×4.5cm，TSS12.0%，酸1.0%。地栽果较大，大小为6.9cm×4.8cm，TSS14.6%。成熟期12月上旬，可树上挂果至春节后。是广东盆栽观赏柑橘的主栽品种。

(5) **Zhushaju** Zhushaju is originated and mainly grown in Guangdong. There is limited cultivation in Guangxi.

Tree spheroid, branches slender and dense. Leaf 7.2~8.0cm×3.8~4.2cm, ovate. Complete flower, small. Fruit obloid, deep red, glossy, size 6.5~6.9cm×4.5~4.8cm, apex and base slightly depressed, easily peeling. TSS 12.0%~14.6%, acid content 1.0%, 15~20 seeds per fruit, taste sour-sweet. Matures in early December. Fruit can hold on tree well after Spring Festival. It is the principal variety of potted ornamental citrus in Guangdong.

25. 紫金春甜橘

来源与分布：紫金春甜橘由广东紫金县选出，2003年通过广东省农作物品种审定委员会认定。主产广东河源、惠州等市，韶关、四会、云浮等市有少量栽培。浙江等省有引种。

主要性状：树势健壮，树冠呈圆头形，枝细长而密。春梢叶片大小为6.5～7.0cm×2.4～2.8cm，阔披针形，叶缘浅锯齿状，无明显翼叶。花较小，完全花。果实扁圆形，橙黄色有光泽，大小为4.8～5.1cm×3.5～3.6cm，果顶平，微凹，果蒂微凸，果皮易剥离。TSS 12.0%～13.4%，酸0.4%～0.5%，果肉清甜较化渣。成熟期2月下旬至3月上旬。该品种早结丰产稳产性好，单一品种栽植属无核橘。主要砧木为三湖红橘、酸橘等。

25. Zijinchuntianju

Origin and Distribution: Selected from Zijin, Guangdong. Zijinchuntianju was registered as a new variety by Guangdong Crop Cultivar Registration Committee in 2003. Its production areas are concentrated in Guangdong Province, mainly in Heyuan and Huizhou, and some in Shaoguan, Sihui, and Yunfu.There are introductions in Zhejiang and other provinces.

Main Characters: Tree vigorous, spheroid, branches long, slender and dense. Leaf size 6.5~7.0cm×2.4~2.8cm , broad lanceolate, margin dentate shallow, petiole wing inconspicuous. Complete flower, comparatively small. Fruit obloid, surface glossy, size 4.8~5.14cm×3.5~3.6cm, apex truncate, slightly depressed, base slightly convex, easily peeling. TSS 12.0%~13.4%, acid content 0.4%~0.5%, pulp pure sweet and relatively melting. Maturity period ranges from late February to early March. The variety is early-fruiting, quite productive. When grown singly, it becomes seedless. The main rootstocks used are Sanhuhongju and Sour tangerine.

（二）柑类 Mandarins

1. 贡柑

来源与分布：原产广东四会，现主产肇庆市的德庆、四会等，清远、云浮、英德、惠州、河源等市有少量栽培，其他省有少量引进。

主要性状：长势中等，树冠圆头形，枝条纤细。春梢叶片大小为6.3～8.8cm×2.8～4.0cm，卵圆形，叶翼较小。花大小介于橙、橘之间，完全花。果实高扁圆形，橙黄色，大小为5.8～6.8cm×5.3～5.9cm，果顶平，微凹。果皮薄紧贴果肉，尚易剥离。TSS 12.0%，酸0.3%～0.5%。果肉脆嫩，汁多化渣，清甜，品质优。种子7粒/果。成熟期12月上中旬。该品种丰产性好，主要砧木为酸橘、三湖红橘、枳等。是我国宽皮柑橘主栽品种之一。

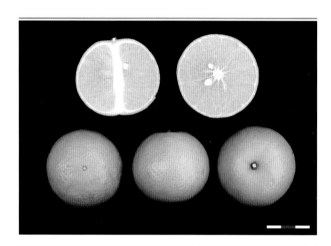

1. Gonggan

Origin and Distribution: Gonggan originated in Sihui, Guangdong Province and is now mainly grown in Deqing and Sihui of Zhaoqing, Guangdong. There are small cultivations in Qingyuan, Yunfu, Yingde, Huizhou and Heyuan of Guangdong Province, and other provinces had introduced it.

Main Characters: Tree medium-vigorous, spheriod, branches slender. Leaf 6.3~8.8cm×2.8~4.0cm in size, ovate, petiole wing relatively small. Complete flower, sized between flowers of orange and mandarin. Fruit long obloid, orange-yellow, 5.8~6.8cm×5.3~5.9cm. Apex truncate, slightly depressed. Rind thin and tightly adherent but still peelable. TSS 12.0%, acid content 0.3%~0.5%. Pulp crisp, juicy and melting, fresh sweet flavor, excellent quality. 7 seeds per fruit. Matures in early to mid-December. The variety is very productive and is one of the most extensively grown loose-skin mandarins in China. Its common rootstocks are Sour tangerine, Sanhuhongju and Trifoliate orange.

少核贡柑 少核贡柑是贡柑的优良变异，2003年在广东省云浮市云安县果园发现，2013年通过广东省品种登记。

树势中等，树冠半圆形，枝条直立，偶有短刺。果实近圆球形，果面光滑，果皮橙黄至橙红色，单果重110.3g，果实大小为6.0cm×5.6cm，果形指数0.93，果皮厚0.15cm；种子4.8粒/果，TSS 11.4%，总糖10.4%，酸0.4%，每100ml果汁维生素C含量25.4mg，果肉爽脆、清甜化渣、少核质优，适应性强。早结丰产，在广东成熟期12月。适宜在广东、广西年有效积温6 000℃以上的地区种植。

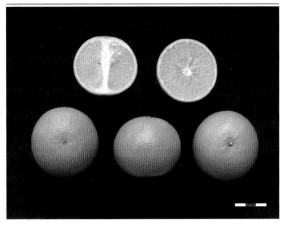

Shaohe Gonggan Shaohe Gonggan is a bud mutation of Gonggan, found in Yun'an County, Yunfu of Guangdong Province and got registration in 2013.

Tree medium in vigor, semi-spheroid, branches upright, short thorns occasionally present. Fruit nearly spherical, rind smooth, orange-yellow to orange-red; Fruit weight 110.3g, 6.0cm × 5.6cm, fruit shape index 0.93, peel 0.15cm thick, 4.8 seeds per fruit, TSS 11.4%, total sugar content 10.4%, TA 0.4%, Vitamin C 25.4mg/100ml. Pulp crisp with good mastication, fresh sweet. Excellent quality and good adaptability. Precocious and productive, maturity in December in Guangdong. It is suitable for mild areas of Guangdong and Guangxi where the annual effective accumulation temperature is above 6 000℃.

2. 黄果柑

来源与分布： 黄果柑又名黄果、广柑、泡皮黄果。原产四川，主产四川南部，贵州、云南有栽培。

主要性状： 树势强健，树冠圆头形，树姿开张，幼树稍直立。春梢叶片大小为8.7～9.2cm×3.5～4.0cm，长卵圆形，叶缘全缘，翼叶剑形。花中等大，完全花。果实近圆形，橙黄色至橙红色，大小为4.0～4.5cm×4.0～4.3cm，果顶微凹，果蒂部凹入明显，有放射沟纹，果皮易剥离。TSS 10.8%～12.0%，酸0.8%～1.1%。无核，品质中等。成熟期1月中下旬。该品种适应性强，耐寒，丰产，果实较耐贮藏。

2. Huangguogan

Origin and Distribution: Huangguogan, also named as Huangguo, Guanggan, Paopihuangguo, is originated in Sichuan and mainly grown in southern Sichuan. There is some cultivation in Guizhou and Yunnan.

Main Characters: Tree vigorous, spheroid topped, spreading, young tree somewhat erect. Leaf 8.7~9.2cm×3.5~4.0cm, long-ovate, margin entire, petiole wing sword-shaped. Complete flower, medium sized. Fruit subglobose, yellow-orange to red-orange, 4.0~4.5cm×4.0~4.3cm in size. Apex slightly depressed; base depressed more pronouncedly, striped with radial grooves. Easily peeling. TSS 10.8%~12.0%, acid content 0.8%～1.1%, seedless, of medium quality. Matures in middle to late January. The tree is widely adaptable, cold-hardy, productive and stores relatively well.

3. 蕉柑

来源与分布：蕉柑又名招柑、桶柑，原产广东潮汕，主要产区是广东的揭阳、潮州、潮阳等市，福建、广西、台湾、四川、贵州有少量栽培。

主要性状：树势中等，树冠圆头形或半圆头形，枝条开张略下垂。春梢叶片大小为4.2～7.5cm×2.1～3.5cm，长椭圆形，两端渐尖。花中大，完全花。果实圆球形或高扁圆形，橙黄至橙红色，果皮稍厚紧贴果肉，尚易剥离，大小为5.7～7.7cm×5.1～6.4cm，果顶平，有的有印圈。TSS 11.0%～14.0%，酸0.4%～0.9%，种子无或少。成熟期12月下旬至翌年1月。该品种早结丰产性好，果肉柔软多汁、化渣、风味浓，有香气。果实耐贮运。主要砧木为酸橘、三湖红橘、红檬檬等。是我国宽皮柑橘主栽品种之一。

3. Jiaogan (Tankan)

Origin and Distribution: Also named as Zhaogan and Tonggan, Jiaogan originated in Chaoshan, Guangdong Province and is mainly grown in Jieyang, Chaozhou and Chaoyang of Guangdong Province. There is small production in Fujian, Guangxi, Taiwan, Sichuan, Guizhou.

Main Characters: Tree medium-vigorous, spheroid or semi-spheroid, branches spreading and somewhat drooping. Leaf on spring shoot 4.2~7.5cm×2.1~3.5cm, long-elliptic, with acuminated apex and base. Complete flower, medium sized. Fruit spheroid or high obloid; yellow-orange to red-orange; rind relatively thick, tightly adherent, peelable. Fruit 5.7~7.7cm×5.1~6.4cm; apex flattened, sometimes marked with areole ring. TSS 11.0%~14.0%, acid content 0.4%~0.9%, seeds few or seedless. Matures from late December to next January. This variety is precocious, very productive. Pulp tender, juicy, melting and rich flavor with pleasant aroma. Stores and ships well. Its main rootstocks are Sour Tangerine, Sanhu-hongju and Red limonia etc. It is one of the major loose-skin mandarins in China.

(1) 白1号蕉柑　广东省普宁市选出的蕉柑新品系。

果实圆球形，橙红色，果面较光滑，果肩不平，大小为5.7cm×5.1cm。TSS 12.0%，酸0.6%，种子1.2粒/果，果肉汁多，甜酸适中，品质上中。成熟期12月下旬至翌年1月上旬。

(1) Bai No.1 Jiaogan　A new line selected from Puning, Guangdong.

Fruit spheroid, 5.7cm×5.1cm, orange-red, surface smooth, base oblique. TSS 12.0%, acid content 0.6%, 1.2 seeds per fruit, pulp juicy, moderate sweet sour, quality medium or above. Maturity period ranges from late December to early January.

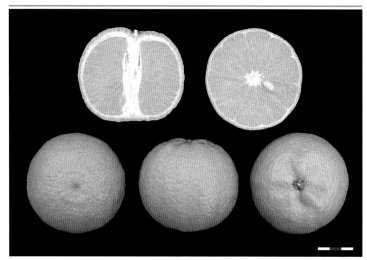

（2）**孚优选蕉柑** 广东潮州市从孚中选蕉柑选育出的新品系。

果实高扁圆形，橙红色，大小为7.7cm×6.7cm。TSS 13.0%，酸0.9%，种子0.3粒/果。果大型端正，肉脆化渣，风味浓，品质上等，成熟期1月中旬。

(2) **Fuyouxuan Jiaogan** A new line selected from Fuzhongxuan Jiaogan in Chaozhou, Guangdong.

Fruit high-obloid, orange-red, 7.7cm×6.7cm in size, TSS 13.0%, acid content 0.9%, 0.3 seeds per fruit. Fruit large, good-looking shape. Pulp crisp and melting, rich in flavor, superior quality. Matures in middle November.

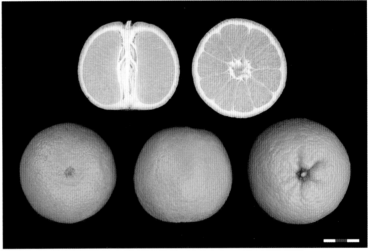

（3）南3号蕉柑　广东普宁市选出的蕉柑新品系。

果实扁圆端正，橙红色，大小为7.0cm×5.6cm，果面油胞突出，果顶平，有不明显印圆。TSS 10.0%～13.0%，酸0.8%，种子2粒/果，品质中上。成熟期12月中下旬。该品系早结丰产性好，较早熟，12月下旬上市。

(3) Nan No.3 Jiaogan　A new line selected from Puning, Guangdong. Fruit obloid, symmetric, 7.0cm×5.6cm in size, orange-red, oil glands conspicuous, apex truncate with faint areole ring. TSS 10.0%~13.0%, acid content 0.8%, 2 seeds per fruit, quality medium or superior. Matures in middle to late December. The variety is precocious, very productive, and relatively early-ripening. It can be marketable in late December.

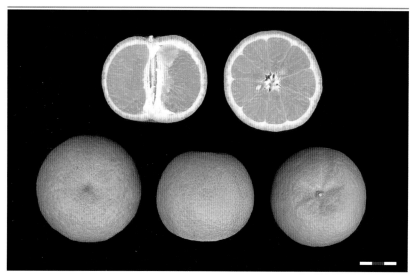

（4）**塔59蕉柑** 广东杨村华侨柑橘场从蕉柑实生树中选出的蕉柑新品系。

果实扁圆形，果面光滑，果蒂有明显的放射沟，橙红色，大小为 5.9～8.6cm×5.1～6.8cm。TSS 12.2%～13.8%，酸0.9%，无籽，品质优。成熟期1月下旬。

(4) Ta 59 Jiaogan A new line selected from seedlings of Jiaogan in Huaqiao Citrus Farm, Yangcun, Guangdong.

Fruit obloid, 5.9~8.6cm× 5.1~6.8cm, surface smooth, orange-red, base striped with conspicuous radial furrows. TSS 12.2%~13.8%, acid content 0.9%, seedless, quality excellent. Matures in late November.

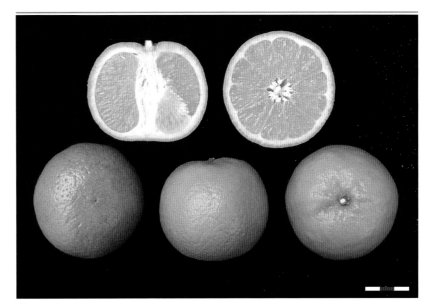

（5）**新1号蕉柑**　广东汕头柑橘研究所选出的蕉柑新品系。

果实高扁圆形，橙红色，较光滑，大小为6.1cm×5.3cm，种子1粒以下。TSS 13.0%，酸1.0%，肉质较嫩，化渣，风味浓，品质上等。成熟期12月下旬至翌年1月。

(5) **Xin No.1 Jiaogan**　A new line selected by Citrus Institute of Shantou, Guangdong.

Fruit long obloid, orange-red; surface moderately smooth. Size 6.1cm×5.3cm, 1 seed to seedless, TSS 13.0%, acid content 1.0%. Pulp crisp and melting, rich in flavor, superior quality. Maturity period ranges from late December to next January.

(6) **粤丰蕉柑** 广东省农业科学院果树研究所等单位从台湾省引进的材料中通过系统选育选出的蕉柑新品系，2006年通过广东省农作物品种审定委员会审定，为广东省重点推广的柑橘新品种。

果实高扁圆形，橙红色，果蒂平，果顶平且有不明显的印环，大小为7.6cm×6.5cm，种子0.6粒/果。TSS 10.5%～12.0%，酸0.8%，肉质较嫩、化渣，品质上等。成熟期1月中下旬。

(6) Yuefeng Jiaogan It is a new line systematically selected from materials introduced from Taiwan, mainly by the Institute of Fruit Tree Research, Guangdong Academy of Agricultural Sciences. It was registered as a new variety by Guangdong Crop Cultivar Registration Committee in 2006 and is the most highly recommended new variety in Guangdong.

Fruit high obloid, 7.6cm×6.5cm, orange-red; base truncate; apex flat with faint areole ring; 0.6 seeds per fruit. TSS 10.5%~12.0%, acid 0.8%. Pulp comparatively crisp and melting; superior quality. Matures in middle to late January.

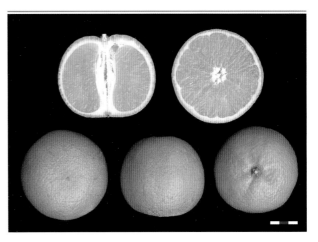

（7）**早熟蕉柑** 广东省农业科学院果树研究所从河婆柑中选出的蕉柑新品系。

果实圆球形，橙红色，大小为5.6～6.2cm×5.4～6.5cm，果蒂微凸，有放射沟纹，果皮薄，光滑。TSS 10.0%～11.0%，酸0.8%，品质中上。成熟期11月上旬，比普通蕉柑早熟40～50d。

(7) Zaoshu Jiaogan (Early-ripening Jiaogan)　A new line selected from Hepogan by the Institute of Fruit Tree Research, Guangdong Academy of Agricultural Sciences.

Fruit globose, 5.6~6.2cm×5.4~6.5cm, orange-red, base slightly depressed and striped with radial grooves, rind thin and smooth. TSS 10.0%~11.0%, acid content 0.8%, quality medium or higher. Matures in early November, 40~50 days earlier than common Jiaogan.

4. 瓯柑

来源与分布：原产浙江温州，是中国古老品种，已有1 000多年栽培历史，主产浙江，福建、广西等地有少量栽培。

主要性状：树势强健，树冠圆头形，枝条开张，下垂，有短刺。春梢叶片大小为6.5～7.0cm×3.2～3.5cm，椭圆形或倒卵形，叶缘锯齿细而不明显，翼叶线状。花大，完全花。果实圆球形或短圆锥形，橙黄色，大小为5.2～5.7cm×5.5～5.8cm，皮肉紧贴，较难剥离。TSS 10.0%左右，酸0.5%～0.6%，种子4.5粒/果。成熟期11月。果实耐贮藏，丰产性好。味酸甜，品质中等，但带有苦味。近年浙江已选出无籽瓯柑新品种。

4. Ougan

Origin and Distribution: Originated in Wenzhou, Zhejiang Province. It is an old variety with a more than 1 000 year-long history of cultivation and now mainly grown in Zhejiang and less in Fujian and Guangxi.

Main Characters: Tree vigorous, crown spheroid, branches spreading and drooping, with short spines. Leaf 6.5~7.0cm×3.2~3.5cm, elliptic or obovate, margin faintly dentate, petiole wing linear. Complete flower, large. Fruit spheroid to short coniform, orange-yellow, 5.5~5.8cm×5.2~5.7cm. Rind tightly adherent, moderately difficult to peel. TSS about 10.0%, acid 0.5%~0.6%. 4.5 seeds per fruit. Tastes sweet and sour with traces of bitterness, quality medium. The variety is productive and stores well. New seedless cultivar has been selected recently in Zhejiang.

5. 温州蜜柑

来源与分布：温州蜜柑原产浙江黄岩，500多年前日本从我国黄岩引进后，从实生变异中选出。日本人以该种原产中国温州，故名温州蜜柑，之后又从芽变中选出很多新品系。我国引进最早是台湾、浙江等省，目前以湖南、浙江、湖北、四川、重庆、广西、江西栽培最多，广东、福建、陕西、云南、贵州、台湾也较多。江苏、安徽有少量栽培。

主要性状：特早熟、早熟品系，树势较弱，树体较小；中熟品系树体中等；晚熟品系树势较壮，树体较大。树冠呈不整齐圆头形或半圆头形，树姿较开张。叶片较大，长椭圆形，各品系间大小不一。花大，单生或丛生，雄性不育。果实扁圆形至倒卵状扁圆形或高扁圆形，果皮易剥离，单果重为120～170g，TSS 9.0%～13.0%，酸0.4%～1.0%，无核，品质佳。成熟期8月上旬至12月中旬。该品种早结丰产性好，适应性强，耐贮藏，适于鲜食和加工。主要砧木为枳和本地早。

5. Satsuma Mandarin

Origin and Distribution: Originated in Wenzhou, Zhejiang Province, China. Satsuma mandarin is believed to be a chance seedling by Japanese researchers from Wenzhou local citrus cultivars brought into Japan about 500 years ago. Because of origin, it was therefore named as unshu mikan (Wenzhou mikan). Now, many new varieties have been developed from long-term clonal selection, such as from bud sport. Modern Satsuma mandarin was first introduced to China via Taiwan and Zhejiang. Currently it is extensively planted in Hunan, Zhejiang, Hubei, Sichuan, Chongqing, Guangxi, and Jiangxi. Guangdong, Fujian, Shaanxi, Yuanan, Guizhou and Taiwan plant it less widely. Jiangsu and Anhui have a small-scale cultivation.

Main Characters: Trees of the early and very early ripening Satsuma mandarins are less vigorous while middle season varieties are moderately vigorous. The late maturity Satsuma mandarins have more vigorous and larger trees. Crown irregular spheroid or semi-spheroid, somewhat spreading. Leaves

relatively large, long-obloid; leaf size varies among cultivars. Flowers large, solitary or in cluster. Stamen sterile. Fruit obloid to obovoid or subglobose, easily peeling; fruit wight 120~170g; TSS 9.0%~13.0%, acid content 0.4%~1.0%, seedless, quality excellent. Matures from early August to middle December. Satsuma mandarins fruit early and are very productive, very adaptable. Fruit stores well and can be used for both fresh and processed consumption. The prevailing rootstocks used for Satsuma mandarins are Trifoliate orange and Bendizao.

Ⅰ. 特早熟温州蜜柑 the Very Early-ripening

（1）大分　从日本引进，为日本大分县柑橘试验场从今田早生温州蜜柑与八朔杂交获得的珠心胚实生选育而成。

树势中强，开张，枝梢节间稍长。果实扁圆形，单果重80～150g，果顶微凹，油胞略凸出，果皮橘黄色，果肉橘红色、化渣。果实8月底开始着色，9月中下旬完全成熟。9月下旬果实TSS10.9%，酸低于1.0%，降酸快。适宜湖北、湖南、浙江等年均气温15℃以上、有效年积温（≥10℃）5 000～6 000℃的地区种植。

(1) Oita　Oita is introduced from Japan, and selected from nucellar embryos of the cross between Imada Wase Satsuma Mandarin and Hassaku Tangor.

Tree medium vigorous, spreading, internodes longer than common Satsuma. Fruit obloid, weight 80~150g, apex slightly depressed oil glands slightly immersed, rind orange-yellow, flesh orange-red and melting. Color change at the end of August and full maturity in mid-late September with TSS 10.9%, TA 1.0% and fruit acid content rapidly reduced to less than 1.0%. It is suitable for Hubei, Hunan, Zhejiang and other areas where the average annual temperature is above 15℃ and the effective annual accumulated temperature (≥10℃) ranges from 5 000~6 000 ℃.

（2）稻叶　日本从早熟宫川温州蜜柑中选出。

果实扁平，橙黄色，大小为7.0cm×5.7cm，果顶微凹，TSS 10.0%，酸1.0%。9月下旬可上市，10月上旬风味最佳。

(2) Inaba Wase　It was selected from early-maturity Miyagawa wase in Japan.

Fruit obloid, yellowish-orange; size 7.0cm×5.7cm; apex slightly depressed; TSS 10.0%, acid content 1.0%. It is marketable in middle -late September. Taste is best in early October.

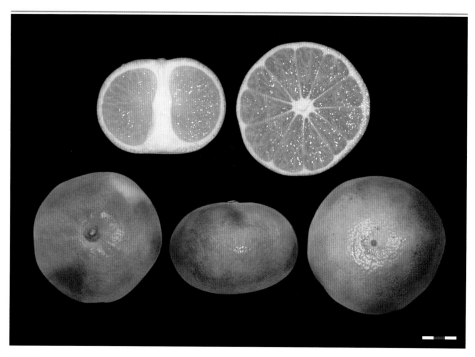

（3）**肥之曙** 从日本引进，1983年由日本研究者用川野夏橙授粉的楠本早生温州蜜柑种子，经珠心胚实生育成。适宜在温州蜜柑适栽区种植。

树势强，幼树直立，结果后开张，枝梢抽发量大，树冠成形快，进入丰产期快。以春梢、秋梢为主要结果母枝，花芽易形成。果实扁圆形，单果重120g，果皮光滑、易剥离。果肉橙红色，囊衣薄，化渣，汁多，风味佳。TSS10.2%，酸0.9%，每100ml果汁维生素C含量26.1mg，可食率77.9%。

(3) Koenoakebono Introduced from Japan, selected from a nucellar seedling of Kusumoto Wase early satsuma mandarin pollinated with Natsumikan in 1983. It is suitable for satsuma mandarin production areas.

Tree vigorous, upright at youth, spreading after fruiting, branching capacity strong, canopy forms quickly, entering fertile period fast. Fruit twigs are mainly spring and autumn shoots, and flower buds easily form. Fruit obloid, rind smooth, easy to peel; flesh orange-red, segment membrane thin, melting, juicy and flavor excellent. Fruit weight 120g, edible portion 77.9%, TSS 10.2%, TA 0.9%, and Vitamin C 26.1mg/100ml.

(4) 宫本 日本从宫川蜜柑变异枝选育而成。

果实扁平，大小一致，橙黄色，大小为6.8cm×5.5cm，果顶平，油胞稍凸起。9月中下旬可上市。TSS 8.0%以上，含酸1.0%以下。积温较高地区较宫川早15d以上。

(4) Miyamoto Wase The variety was selected from limb sport of Miyagawa wase in Japan.

Fruit obloid, yellowish orange, sizes uniformly in 6.8cm×5.5cm, apex flattened, oil glands slightly conspicuous. Fruit marketing season commences in middle-late September. TSS above 8.0%, acid content below 1.0%. In the areas with higher accumulated temperature, It matures 15 days earlier than Miyagawa wase.

（5）**国庆1号** 华中农业大学从龟井温州蜜柑的变异中选出。

果实扁圆或高扁圆形，橙黄至橙色，大小为5.9cm×4.8cm，果肉橙红色，味浓甜，TSS 11.5%，酸0.6%，10月上旬成熟。

(5) Guoqing No.1 A selection from Kamei unshu by Huazhong Agricultural University.

Fruit obloid to longobloid, yellow-orange to orange, size 5.9cm×4.8cm; flesh red-orange; flavor rich sweet; TSS 11.5%, acid content 0.6%; matures in early October.

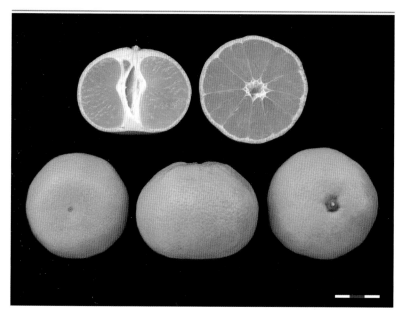

(6)隆园早　1973年隆回县园艺场从松木早熟温州蜜柑的变异中选出。

果实扁圆形，橙黄或深橙黄，大小为6.8cm×5.2cm，果顶常有小脐。果肉脆嫩，风味甜酸较浓，化渣，TSS 11.3%～12.9%，酸0.64%～1.0%。9月中旬着色，9月底成熟。

(6) Longyuanzao　A mutant of early-maturity Matsuki discovered in the orchard of Longhui County in 1973.

Fruit obloid, yellow-orange to deep orange-yellow, size 6.8cm×5.2cm, apex often contains small navel. Flesh crisp, flavor rich sweet-sour, melting, TSS 11.3%~12.9%, acid content 0.64%~1.0%. Color breaks in mid-September and fruit matures in late September.

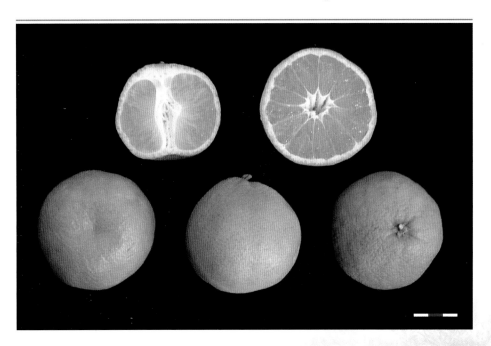

（7）日南1号 日本从兴津早熟温州蜜柑变异中选出。

果实扁圆，橙黄色，大小为6.8 cm×4.8 cm，果顶平，果蒂部微凹。10月上旬TSS 11.0%，酸1.0%以下可上市。

(7) Nichinan No.1 Wase   A selection from mutations of early-maturity Okitsu wase in Japan. Fruit obloid, yellowish-orange; size 6.8cm × 4.8cm; apex flattened; base slightly depressed. Matures in early October. It is marketable when TSS amounts to 11.0% and acid content reduces to 1.0%.

(8) **市文**　日本从宫川温州蜜柑中选出。

果实扁平，是特早熟种中最扁的种。橙黄色，大小为6.9cm×4.8cm，果顶平。9月下旬可上市，TSS 9.0%～10.0%，含酸1.0%以下，9月下旬至10月上旬风味最佳。

(8) Ichifumi Wase　The variety was selected from Miyagawa wase in Japan.

Fruit obloid (actually the most depressed among the extremely early Satsuma varieties), yellowish-orange; size 6.9cm×4.8cm. Apex flattened. Fruit is marketable in late September when TSS is 9.0%~10.0% and acid content is below 1.0%. Best flavor is reached in late September and early October.

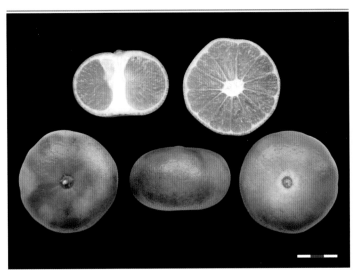

（9）**由良**　引自日本，由宫川芽变而来，2016年通过浙江省林木良种认定。主产浙江，国内柑橘主要产区均有引种试种。

树势中强，树冠自然开心形。果实扁圆形或高扁圆形，单果重100g，TSS12.0%～15.0%，酸0.8%～1.0%，糖度高，囊衣薄，化渣性好，风味浓郁。成熟期10月上旬左右，可在国庆前后上市。丰产稳产，栽培容易。成熟后雨水对降酸与果形有一定影响。积温不超过6 000℃的柑橘适栽区均可种植。

(9) **Yura Wase**　It was introduced from Japan, selected from Miyagawa bud sport and registered in Zhejiang Province in 2016. It is mainly produced in Zhejiang Province, and has also been introduced to other citrus production areas of China for trial.

Tree medium-vigorous, crown naturally open-center. Fruit obloid or long obloid. Single fruit weight about 100g with TSS 12.0%~15.0%, TA 0.8%~1.0%; segment membrane thin, pulp melting, high in sweetness, flavor rich. The maturity period is early October and meets the holiday market supply for Chinese National Day. The variety is easy to grow with a consistent and high yield. Rainfall after ripening has a certain effect on acid reduction and fruit shape.

It is suitable for citrus production regions where the accumulated temperature is less than 6 000℃.

II. 早熟温州蜜柑 the Early-ripening

（1）**鄂柑2号** 鄂柑2号又名光明早，是湖北宜都市和华中农业大学在吴家乡光明村从龟井早熟温州蜜柑园中选出。

果实扁圆形，单果重130g左右，果顶平微凹。TSS 11.4%，酸0.98%。味浓，有微香。成熟期10月上旬。该品种丰产稳产，耐寒，北缘产区种植表现良好。

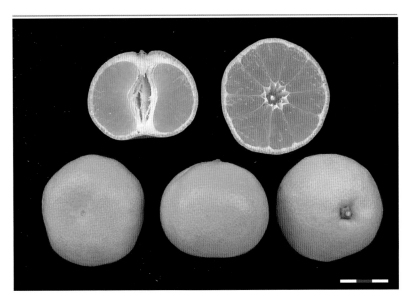

(1) E-gan No.2 Also named as Guangmingzao, E-gan No.2 is selected from Kamei unshu orchard of Guangming village, Wujia town, Yichang Hubei, China by Yidu County of Hubei and Huazhong Agricultural University.

Fruit obloid, weight about 130g, apex flattened and slightly depressed. TSS 11.4%, acid content 0.98%. Flavor rich with weak aroma. Matures in early October. Tree yields consistently and very productive, cold-hardy. Well performance in northern citrus growing regions.

（2）宫川　日本从温州蜜柑的芽变中选出。

果实高扁圆形，果面光滑，橙黄至橙色，大小为6.7cm×5.1cm，果蒂部略凸，有4～5条放射沟。TSS 10.0%～12.0%，酸0.6%～0.7%。该品种早结丰产，果形整齐美观，品质优。成熟期10月中旬。是我国早熟温州蜜柑的主栽品种。

(2) Miyagawa Wase　It is a bud mutation selected from Satsuma mandarin in Japan.

Fruit long obloid, rind smooth, yellow-orange to orange, size 6.7cm×5.1cm, base slightly convex with 4~5 radial furrows. TSS 10.0%~12.0%, acid content 0.6%~0.7%. Early fruiting and productive. Fruits are uniform in size and beautiful in appearance, quality excellent. Matures in middle October. Miyagawa wase is the principle variety of early-ripening Satsuma mandarins cultivated in China.

（3）立间　日本从尾张温州蜜柑的变异中选出。

果实扁圆形或近球形，橙色，较光滑，大小为6.7cm×5.4cm。果顶圆钝，果皮光滑，TSS 10.0%～11.0%，酸0.5%～0.6%。该品种早结丰产，品质好，有微香。树体较矮小，可适当密植，成熟期10月上中旬。

(3) Tachima Wase　It originated as a mutation of Owari unshu in Japan.

Fruit obloid or subglobose, rind orange-colored, smooth, 6.7cm×5.4cm. Apex obtusely rounded, TSS 10.0%~11.0%, acid content 0.5%~0.6%. The variety is precocious and prolific. Fruit has weak aroma and excellent quality. Trees moderately dwarf, and can be planted in a modcratcly higher density. Matures in early-middle October.

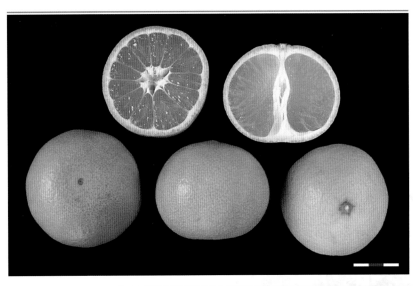

（4）兴津　日本兴津园艺场以宫川为母本、枳为父本杂交后代的珠心苗中选出。

果实扁圆形，橙红色，大小为7.0cm×5.7cm，果顶圆或平圆，TSS 10.0%～11.0%，酸0.7%。该品种早结丰产，适应性广，品质优，不易裂果，成熟期10月上旬。是我国早熟温州蜜柑的主栽品种。

(4) Okitsu Wase   Okitsu wase is a nucellar seedling from a cross of Miyagawa wase (female parent) and *P. trifoliata* (male parent), which was selected at the Horticultural Research Station, Okitsu, Japan.

Fruit obloid, orange-red, size 7.0cm×5.7cm, apex rounded or flattened, TSS 10.0%~11.0%, acid content 0.7%. Tree is early-fruiting and prolific, widely adaptable. Fruit quality is excellent without significant splitting problem. Matures in early October. It is one of the main varieties of early-ripening Satsuma mandarins in China.

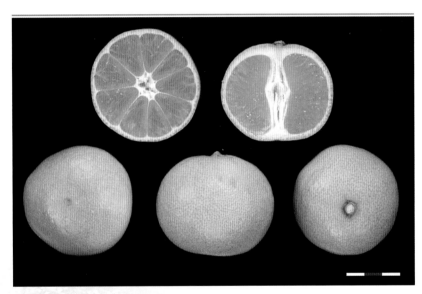

III. 中熟与晚熟温州蜜柑　the Middle- and Late-ripening

(1) 涟红　湖南涟源县园艺场从尾张温州蜜柑的变异中选出。主产湖南。

果实近扁圆形，橙红色，大小为6.5cm×5.0cm，果顶微凹。TSS 10.5%～12.4%，酸0.6%。成熟期10月下旬至11月上旬。该品种早结丰产，品质优，适宜鲜食和加工糖水橘瓣罐头。

(1) Lianhong   It was selected from a variation of Owari unshu in an orchard of Lianyuan County, Hunan. The main growing region is in Hunan.

Fruit obloid, orange-red, 6.5cm×5.0cm in size, apex slightly depressed. TSS 10.5%~12.4%, acid content 0.6%. Matures from late October to early November. The variety is precocious and productive with excellent quality. It fits for both fresh and canned consumption.

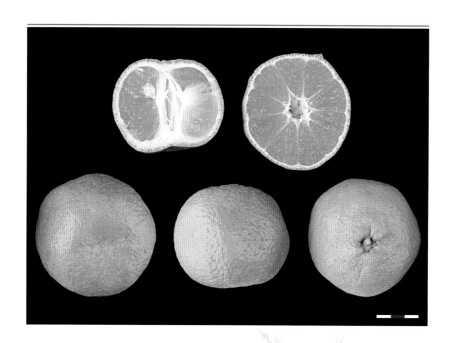

(2)**南柑20** 日本从尾张温州蜜柑芽变中选出。

果实扁圆形，深橙色，大小为6.5cm×5.0cm，果顶平而圆钝。TSS 11.0%～13.0%，酸0.6%～0.7%，风味浓，成熟期10月下旬至11月上旬。该品种中熟偏早，适应性强。

(2) Nankan No.20 It is a bud mutation derived from Owari unshu in Japan.

Fruit obloid, deep orange, size 6.5cm×5.0cm. Apex flattened and obtusely rounded. TSS 11.0%~13.0%, acid content 0.6%~0.7%, rich flavor. Matures in late October to early November. The variety is middle to slightly early in maturity and is very adaptable.

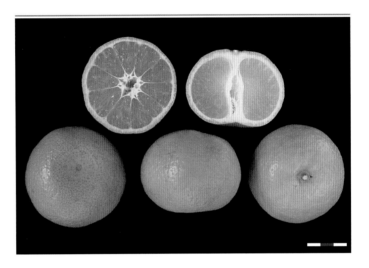

(3)尾张　日本从伊木力系的变异中选出。

果实扁圆形，橙黄色，大小为6.7cm×5.1cm，果顶印圈明显。TSS 11.5%～12.0%，酸0.5%～1.0%。果肉细嫩，味甜酸，品质中等。成熟期11月中下旬。该品种丰产性好，不易裂果，是我国中熟温州蜜柑的主栽品种。

(3) Owari　It was selected from mutants of Ikiriki in Japan.

Fruit obloid, yellow-orange, 6.7cm×5.1cm in size, areole ring prominent. TSS 11.5%~12.0%, acid content 0.5%~1.0%. Flesh tender, sweet sour taste, medium quality. Matures in middle to late November. Productive and without significant fruit splitting problem. It is one of the main middle season Satsuma varieties in China.

（三）杂柑类 Hybrids

1. 1232橘橙

来源与分布：中国农业科学院柑橘研究所以成年的伏令夏橙为母本、以江南柑和朱砂柑为父本进行杂交育成，重庆市等地有少量栽培。

主要性状：树势强，枝条较直立，小枝质地坚硬，有短刺。春梢叶片大小为8.0～10.2cm×3.0～3.8cm，尖卵形。果实球形至椭圆形，红色，大小为5.2cm×5.1cm，果顶微凹，有放射沟数条，果皮较易剥离。TSS 11.5%，酸0.83%。成熟期11月下旬至12月中旬。该品种耐贮性好，不加处理常温保存至翌年3～4月仍味甜可口，无汁胞粒化现象。

1. Tangor No.1232

Origin and Distribution: It is a hybrid between adult Valencia orange (female parent) and Jiangnangan and Zhushagan (male parents), selected by the Citrus Research Institute of Chinese Academy of Agricultural Sciences, and cultivated a little in Chongqing.

Main Characters: Tree vigorous, branches comparatively erect, branchlets hard, with short spines. Leaf size 8.0~10.2cm×3.0~3.8cm, pointed ovate. Fruit spheroid to ellipsoid, red, size 5.2cm×5.1cm. Apex slightly depressed, striped with several radial furrows. Peels somewhat readily. TSS 11.5%, acid content 0.83%. Maturity season from late November to Mid-December. Fruit stores well, tastes still good until next March to April under ambient temperature condition without any special treatment and no vesicle granulation appears.

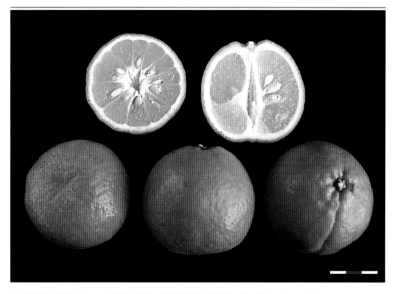

2. 不知火

来源与分布：日本农林水产省用清见和中野3号椪柑杂交育成。我国于1992年从日本引入，浙江、重庆、湖北、广东有少量试种。在我国四川种植比较成功，别名丑柑。

主要性状：树势中等，幼树树姿较直立，进入结果后开张，枝梢密生细而短，刺随树体长大而消失。春梢叶片大小为7.3cm×3.2cm，卵圆形，与椪柑相似。花比清见和椪柑大，多数是畸形花，花粉量少。果实倒卵形，多数有短颈，无短颈的扁果顶部有脐，橙黄色，大小为7.2cm×7.4cm，果皮易剥离。TSS 13.0%～14.0%，酸1.0%，无核。果肉脆、多汁化渣，有椪柑香气，品质佳。12月上旬完成着色，成熟期翌年2～3月。该品种适应性强，栽培区广，最适在冬天无霜冻的中、南亚热带气候区栽培。该品种需要较长时间降酸。以枳或红橘作砧木。

2. Shiranui

Origin and Distribution: Shiranui originated from a cross between Kiyomi and Nakano No.3 ponkan, made by the Ministry of Agriculture, Forestry and Fisheries of Japan. It was introduced to China from Japan in 1992 and is mostly grown in Sichuan. There are limited cultivations in Zhejiang, Chongqing, Hubei and Guangdong. Mainly grown in Sichuan Province, also known as Chougan (ugly mandarin).

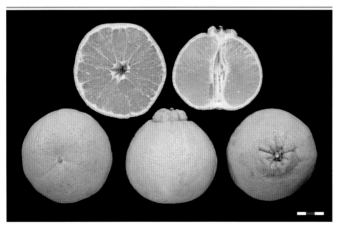

Main Characters: Tree medium-vigorous, young tree somewhat erect, spreading as fruiting progresses. Branches dense, slender and short, spines disappear as the trees grow up. Leaf 7.3cm×3.2cm in size, ovate shape like Ponkan. Flowers larger than Qingjian and Ponkan, mostly malformed with few pollen grains. Fruit obovoid, orange-yellow, 7.2cm×7.4cm, mostly with a low collar at base; collar-less obloid fruit with navel. Easily peeling. TSS 13.0%~14.0%, acid content 1.0%, seedless. Pulp crisp, juicy and melting, with Ponkan-like aroma, excellent quality. Colors completely in early December and matures in February to March. Moderately adaptable in most areas, best to cultivate in frost-free areas of the central and southern semi-tropical regions. It needs a long time to decrease the acidity. Trifoliate orange or Hongju can be used as its rootstocks.

3. 春香

来源与分布：从日本引进，来源于日向夏实生后代。适宜我国长江以南柑橘产区种植。

主要性状：生长初期树势较旺，结果后枝条逐渐开张，生长势中等。初果期果实较大，果实近圆形，果皮粗厚、黄色，果顶有圆形深凹圈印与乳头状凸起。果肉淡黄色，味清甜质优，囊衣厚不化渣，种子少。单果重264.3g，TSS12.0%，可食率70.3%。果实成熟期12月上旬。早结性能好，丰产、稳产。

3. Hiruka Tangor

Origin and Distribution: Introduced from Japan, selected from the chance seedling of *C. tamurana*. It is suitable for citrus production area located in the south of Yangtze River.

Main characters: Tree quite vigorous at the initial stage of growth, medium vigorous after fruiting when branches open gradually. At the early fruiting stage, the fruit is large, nearly spheroid, pericarp thick, and yellow; Apex of a deep round concave imprint and a nipple-like structure. Pulp light yellow, flavor fresh sweet, quality excellent; Segment membrane thick, not melting, a few seeds. Average fruit weight 264.3g, TSS 12.0%, edible portion 70.3%. Fruit maturity in early December. Precocious and stably productive.

4. 大雅柑

来源与分布： 大雅柑为清见橘橙和新生系3号椪柑的杂交后代，中国农业科学院柑橘研究所1998年开展杂交，2016年通过四川省农作物品种审定委员会审定。

主要性状： 树势强，成枝力强，结果后树冠略开张。以单枝顶花结果为主，坐果率高，丰产稳产。果面光滑，深橙色，单果重169.3～212.6g，果实大小7.8～8.4cm×5.8～7.4cm，果形指数0.75～0.90，果皮厚0.34～0.40cm，无核，可食率70%～74%，TSS12.0%以上，总糖11.1%～11.8%，酸0.4%～0.6%，每100ml果汁维生素C含量30.1～35.1mg，固酸比17.1～32.5，综合品质优良，果汁丰富，果肉脆嫩化渣，风味浓郁，口感好，有香气，易剥皮分瓣。晚熟，降酸较春见更晚，采收供应期长，在四川眉山和重庆北碚在地区翌年1月中下旬至3月下旬达到优质商品果采收上市条件。适宜在冬季无冻害、春季回温较慢的区域种植。

4. Dayagan Tangor

Origin and Distribution: Dayagan tangor is an offspring of a hybrid (Kiyomi crossed with ponkan nucellar line No.3), developed by Citrus Research Institute, Chinese Academy of Agricultural Sciences in 1998, and registered as a new cultivar in Sichuan Province in 2016.

Main Characters: Trees vigorous, branching capacity strong, canopy open after fruiting. Mainly single-branch terminal flowers bear fruit. Fruit setting rate high, prolific and yield stable. Fruit surface smooth, dark-orange, fruit weight 169.3~212.6g, size in 7.8~8.4cm×5.8~7.4cm, fruit shape index 0.75~0.90, pericarp thickness 0.34~0.40cm, seedless, edible portion 70%~74%, TSS 12.0%, total sugar content 11.1%~11.8%, TA 0.4%~0.6%, Vitamin C 30.1~35.1mg/100ml juice, TSS/TA ratio (RTT) 17.1~32.5. Over-all quality excellent, juicy, pulp crisp and tender, melting, flavor rich, tasty, fragrant, easy to peel and dissect. Late ripening, fruit acidity decreases later than that of Harumi, harvest and supply period long. Quality peaks when harvested in the following January to March in Meishan, Sichuan province and Beibei, Chongqing. It is suitable for areas in Sichuan without frost damage in winter and slow temperature rising in spring.

5. 甘平

来源与分布：原产日本，西之香×椪柑杂交选育。主产浙江、四川。

主要性状：树势强，幼树较直立，结果后开张。果实扁平，果皮光滑，橙红色，果皮较薄，厚约2.0mm，易剥离，中心柱空，单果重300g，果形指数0.68；果肉糖度高、风味浓，化渣性好，口感佳，TSS13.0%～15.0%，酸0.8%～1.1%，无核。在浙江地区1月下旬至3月上旬成熟；大小年较明显，易裂果、枯水。植株抗寒性中等，耐-5℃低温。适合于冬季低温0℃以上，光照良好地区栽培。冬季低温在0℃以下地区须采取大棚保温栽培。

5. Kanpei

Origin and Distribution: Hybrid of Nishinokaori and Ponkan, introduced from Japan and mainly grown in Zhejiang and Sichuan provinces.

Main Characters: Tree vigorous, young tree upright, and opening after fruiting. Fruit obloid, with a smooth and orange-red peel about 2.0mm thick, easy to peel; Axis empty, single-fruit weight 300g, fruit shape index 0.68; Pulp is high in sugar content and rich flavor, TSS 13.0%~15.0%, TA 0.8%~1.1%, melting and seedless. It matures from late January to early March in Zhejiang. This variety has the drawbacks of alternate bearing, easy fruit cracking and dehydration. Moderate cold resistance up to minus 5℃. It is suitable for cultivation in areas with low temperature above zero and good sunlight in the winter. In the area where the low temperature below 0 ℃, it must be cultivated in film greenhouses.

6. 高橙

来源与分布：原产浙江省温岭，是柚与橙或橘的自然杂种，浙江省有少量栽培。

主要性状：树势强健，树冠自然圆头形，枝梢粗壮，有刺。春梢叶片大小为8.5cm×7.6cm。果实倒卵圆形，橙黄色，大小为9.0～11.0cm×7.5～10.0cm，果面稍粗糙。TSS 11.0%，酸1.5%。种子8～12粒/果。成熟期11月中下旬，也可延迟到翌年1～2月采收。该品种是一个品味独特的地方良种，适应性广，抗逆性强，丰产性好，果实耐贮藏。

6. Gaocheng

Origin and Distribution: Originated in Wenling, Zhejing Province, Gaocheng is a natural cross between pummelo and orange and cultivated a little in Zhejiang.

Main Characters: Tree vigorous, natural spheroid, branches thick and hard, with spines. Leaf 8.5cm×7.6cm. Fruit obovoid, orange-yellow, 9.0~11.0cm×7.5~10.0cm, surface slightly rough. TSS 11.0%, acid content 1.5%, 8~12 seeds per fruit. Matures in middle to late November. Fruits could hold on tree well to the next January and February. It is a fine local variety of distinctive flavor, wide adaptability, strong resistance to stress conditions and high productivity. Fruit stores well.

7. 红锦

来源与分布： 四川省农业科学院园艺研究所采用自选1-15橘橙作母本、红橘作父本杂交育成的三倍体柑橘品种，又称红锦橘橙。2017年获得植物新品种权。四川有少量栽培。

主要性状： 树势强，树冠较开张，萌芽力高，成枝力强。结果母枝以春梢和早秋梢为主，坐果率高，丰产稳产。果实高桩扁圆形，果顶平，果形端正。果皮红色或着红晕，光滑、具光泽，果肉深橙色，果皮较易剥离，单果重123.7~184.9g，大小为6.7~7.4cm×5.4~6.5cm，果形指数0.81~0.91，果皮厚0.30~0.43cm，囊瓣数8~12。TSS 10.0%~13.3%，酸0.4%~1.0%，固酸比11.2~25.2，每100ml果汁维生素C含量31.1~47.2mg，可食率65.2%~73.9%，果汁率51.3%~58.7%，种子1.7粒/果。甜酸适度，化渣，耐贮运。在四川翌年2月成熟。适宜在晚熟柑橘区种植。

7. Hongjin Tangor

Origin and Distribution: This variety is a triploid from the crossing of Tangor 1-15 with Red Tangerine as the male parent, developed by the Horticulture Research Institute of the Sichuan Academy of Agricultural Sciences, and got PVP of the Ministry of Agriculture in 2017. It is now cultivated in Sichuan Province with a small amount of acreage.

Main Characters: Tree vigorous with open canopy, strong budding and branching capacity. The fruiting twigs are mainly spring and early autumn shoots. Prolific and yield stable. Fruits long obloid, apex flattened, fruit shape regular and uniform. Rind smooth, red or glossy red, moderately easy to peel. Fruit weight 123.7~184.9g, size in 6.7~7.4cm×5.4~6.5cm, fruit shape index 0.81~0.91, peel thickness 0.30~0.43cm, segments 8~12, 1.7 seeds in average, edible portion 65.2%~73.9%, juicing rate 51.3%~58.7%; TSS 10.0%~13.3%, TA 0.4%~1.0%, Vitamin C 31.1~47.2mg/100ml. Moderate sweet and sour, melting, stores well. Mature in the following February in Sichuan. Suitable areas for late-season citrus.

8. 红美人

来源与分布：原产日本，南香×天草杂交选育。在四川、浙江、湖北、江西等地栽培。

主要性状：幼树树势强，结果后树势中庸。单果重250g，果实扁圆形至高扁圆形，有脐点；果皮较光滑，红橙色，剥皮难易中等；中心柱小，半空虚；高糖低酸，囊壁薄，柔软多汁，香味浓郁，TSS12.0%～15.0%；无核。对溃疡病较敏感。植株抗寒性中上，耐-7℃低温。适宜设施栽培，或年均温18℃以上、光照良好、采前降雨较少的地区露天栽培。

8. Beni Madonna

Origin and Distribution: Introduced from Japan, selected from the cross of Nankou × Amakusa. Mainly grown in Sichuan, Zhejiang, Hubei, Jiangxi and other provinces.

Main Characters: Tree vigorous at youth and medium vigorous after fruiting; average fruit weight 250g, fruit shape obloid to long obloid, with mini navel; Surface smooth, reddish-orange, moderate difficulty to peel; Axis small and semi-empty; high in sugar and low in acid, thin segment membrane, soft and succulent pulp, rich aroma, TSS 12.0%~15.0%, seedless. Sensitive to citrus canker, medium-to-high cold resistance up to −7 ℃. It is suitable for cultivation in facilities, or in the open area where the annual temperature is above 18 ℃, with good sunlight and few rainfalls in the pre-harvest season.

9. 红柿柑

来源与分布：红柿柑又名439橘橙。浙江省柑橘研究所以瓯柑为母本、以改良橙为父本杂交育成的新品种，重庆等地有少量栽培。

主要性状：树势强健，树冠圆锥形或圆柱形。枝梢密集，个别枝条下垂。春梢叶片大小为8.7cm×4.4cm，披针形或椭圆形。果实圆球形，橙红色，大小为5.2cm×5.1cm，果顶有明显印圈和小脐孔。TSS 14.0%～16.0%，酸1.2%，果汁多，化渣，有香气，品质优良。种子14～16粒/果。成熟期11月下旬。该品种适应性强，抗旱耐瘠性强，丰产稳产。蟹橙砧树冠高大，枸头橙砧树冠较小。

9. Hongshigan

Origin and Distribution: Also named as 439 Jucheng. This new variety originated from a cross between Ougan (female parent) and Gailiangcheng (male parent), made by the Citrus Research Institute of Zhejiang Province, and is cultivated to some extent in Chongqing etc.

Main Characters: Tree vigorous, coniform or columniform. Branches dense, occasionally drooping. Leaf 8.7cm×4.4cm, lanceolate or elliptic. Fruit spheroid, orange-red, 5.2cm×5.1cm. Apex with conspicuous areole ring and a small open navel. TSS 14.0%~16.0%, acid content 1.2%. Pulp juicy, melting, fragrant, of excellent quality. 14~16 seeds per fruit. Matures in late November. The variety is widely adaptable, very tolerant to drought and lean soil. Yields are high and consistent. Trees on Xiecheng rootstock are tall whereas trees on Goutoucheng are smaller.

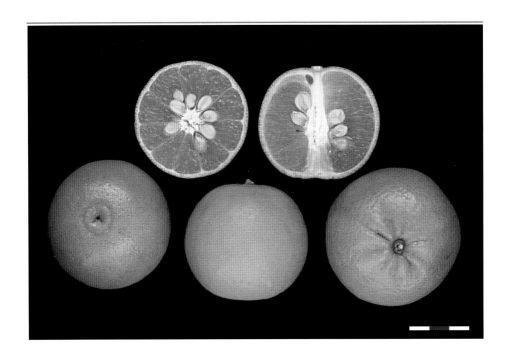

10. 红玉柑

来源与分布：浙江省柑橘研究所用新本1号（少核本地早）与刘本橙（刘金光甜橙×本地早），通过回交途径选育而成。

主要性状：树势强健，树姿开张，树冠呈自然圆头形，枝梢密生，少数夏梢有刺。春梢叶片大小为6.3cm×3.2cm，椭圆状披针形。花期较本地早提早5～7d。果实高扁圆形，橙黄色，大小为7.3cm×5.9cm，皮较紧，可以剥离。TSS 12.5%～14.6%，酸0.8%～1.0%。果肉橙红色，质地脆嫩，汁多。单一品种种植常无核。成熟期11月下旬。该品种丰产、稳产、耐贮藏，常规条件下，贮至翌年4月中旬不枯水、不变味。以枸头橙、蟹橙、本地早等作砧木。

10. Hongyugan

Origin and Distribution: This variety originated from a cross of Xinben No.1 (Shaohebendizao) and Liubencheng (a cross between Liujinguang sweet orange and Bendizao mandarin), made by the Citrus Research Institute of Zhejiang Province.

Main Characters: Tree vigorous, natural spheroid, spreading, branches dense, some summer twigs thorny. Leaf 6.3cm×3.2cm, elliptically lanceolate. Flowering 5~7 days earlier than Bendizao. Fruit high obloid, orange-yellow, 7.3cm×5.9cm in size. Rind relatively tight adherent, peelable. TSS 12.5%~14.6%, acid content 0.8%~1.0%. Pulp orange-red, crisp and tender, juicy. Fruit is seedless when planted in single variety. Matures in late November. Yields are high and consistent. Fruit stores well and remains juicy, fragrant till next April under conventional storage conditions. Commonly used rootstocks are Goutoucheng, Xiecheng and Bendizao etc.

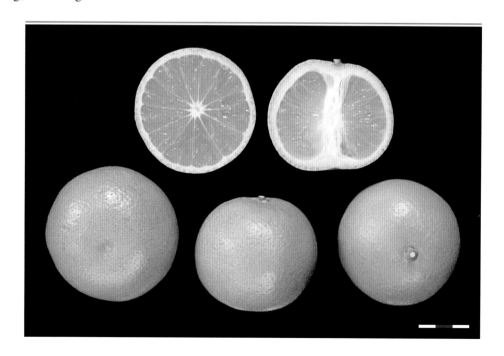

11. 红韵香柑

来源与分布：红韵香柑是塔罗科血橙和W.默科特的杂交后代，橘橙杂柑。西南大学柑橘研究所（中国农业科学院柑橘研究所）2006年开展杂交。2020年获得植物新品种权。

主要性状：树势强，树冠较紧凑，萌芽力强，成枝力强。结果母枝以春梢、早秋梢为主。坐果率高，丰产稳产。果实扁圆形，果面有沙粒状油胞，果皮为橙黄色带花青素着色，具特殊的花椒香气，单果重120g，大小为5.5～6.8cm×5.0～5.5cm，果形指数0.74～0.91，果皮厚0.21～0.43cm，可食率68.7%～71.7%，TSS12.4%～14.5%，总糖10.2%～10.6%，酸0.7%～1.2%，每100ml果汁维生素C含量32.0～36.7mg，固酸比10.3～16.1，果肉呈现少量星点状花青素着色，果肉及果汁有令人愉悦的香气，且香气特别突出，甜酸适度、化渣，品质优良，耐贮性强。种子4粒/果，易剥皮分瓣。在重庆北碚气候条件下，品质在翌年3月达到最优。适宜在冬季冷凉但无冻害的甜橙产区栽培。

11. Hongyun Xianggan Tangor

Origin and Distribution: Hongyun Xianggan is an offspring of the cross between Tarrocco Blood Orange and W. Murcott, hybridized in 2006 by Citrus Research Institute of Chinese Academy of Agricultural Sciences. The variety has got PVP of the Ministry of Agriculture and Rural Affairs since 2020.

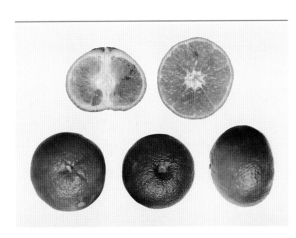

Main Characters: Trees vigorous, crown somewhat compact, budding and branching capacity strong. Mainly spring and early autumn shoots bear fruits. Prolific with high fruit setting rate, yield stable. Fruit quality peaks in the following March in Beibei, Chongqing. Fruit oblate, Rind rough, with oil glands, orange-yellow with anthocyanin pigmentation, of distinctive peppery aroma. Fruit weight 120g, size in 5.5~6.8cm× 5.0~5.5cm, fruit shape index 0.74~0.91, pericarp thickness 0.21~0.43cm, edible portion 68.7%~71.7%, TSS 12.4%~14.5%, total sugar content 10.2%~10.6%, TA 0.7%~1.2%, Vitamin C 32.0~36.7mg/100ml juice, 4 seeds per fruit, easy to peel and dissect. Pulp spotted with anthocyanin pigmentation, both pulp and juice with a pleasant and prominent aroma, flavor moderately sweet and sour, melting. Fruit quality is excellent and stores well.

It is suitable for sweet orange production area with cool winter but no frost damage.

12. 金春

来源与分布：四川省农业科学院园艺研究所采用自选1-15橘橙作母本、普通椪柑作父本杂交育成，又称金春橘橙。2017年获得植物新品种权。四川有少量栽培。

主要性状：树势强，树冠自然圆头形，萌芽力和成枝力强。结果母枝以春梢和早秋梢为主，坐果率高，丰产稳产。果实扁圆形，端正对称，果面光滑，果顶和果蒂平，果皮橙色，易剥离，果肉深橙色。单果重138.4～246.6g，大小为7.4～8.3cm×5.2～7.6cm，果形指数0.76～0.86，果皮厚0.28～0.43cm，囊瓣数10～12。TSS 10.7%～12.6%，酸0.7%～0.9%，固酸比11.7～17.5，每100ml果汁维生素C含量42.0～54.0mg，可食率69.9%～77.8%，果汁率50.5%～59.1%，种子5.5粒/果。甜酸适度、化渣、耐贮运。在四川翌年2～3月成熟。适宜在晚熟柑橘区种植。

12. Jinchun Tangor

Origin and Distribution: This variety is a hybrid from the crossing of Tangor 1-15 with Ponkan as the male parent, developed by the Horticulture Research Institute of the Sichuan Academy of Agricultural Sciences, and got PVP of the Ministry of Agriculture in 2017. It is now grown in Sichuan Province with a small amount of acreage.

Main Characters: Tree vigorous, crown natural spheroid, budding and branching capacity strong. The fruiting twigs are mainly spring and early autumn shoots. Prolific and yield stable. Fruit oblate, symmetrical, with flattened apex and base; Rind smooth, orange in color, easy to peel. Fruit weight 138.4~246.6g, size in 7.4~8.3cm×5.2~7.6cm, fruit shape index 0.76~0.86, peel thickness 0.28~0.43cm, segments 10~12. TSS 10.7%~12.6%, TA 0.7%~0.9%, TSS/TA 11.7~17.5, Vitamin C 42.0~54.0mg/100ml. Edible portion 69.9%~77.8%, juicing rate 50.5%~59.1%, 5.5 seeds per fruit; Moderate sweet and sour, melting, stores well. Maturity in Sichuan is from February to March in the following year. Suitable areas for late-season citrus.

13. 金煌

来源与分布：四川省农业科学院园艺研究所用自选1-15橘橙作母本、普通椪柑作父本杂交育成，又称金煌橘橙。2017年获得植物新品种权。四川有少量栽培。

主要性状：树势强，树冠自然圆头形，萌芽力和成枝力强。结果母枝以春梢和早秋梢为主，坐果率高，丰产稳产。果实倒卵圆形，果面中等粗细，蒂部有短颈，果皮橙色，易剥离，果肉深橙色。单果重236.8～359.8g，大小为8.2～9.4cm×8.2～10.8cm，果形指数0.95～1.23，果皮厚0.30～0.42cm，囊瓣数12～14。TSS 13.3%～16.2%，酸0.4%～0.9%，每100ml果汁维生素C含量31.3～56.6mg。可食率61.4%～71.0%，果汁率42.1%～57.2%，种子4.14粒/果（混栽）。甜酸适度、风味浓、脆嫩化渣，耐贮运。在四川翌年2～3月成熟。适宜在晚熟柑橘区种植。

13. Jinhuang Tangor

Origin and Distribution: This variety is a hybrid of Tangor 1-15 crossed with Ponkan as the male parent, developed by the Horticulture Research Institute of the Sichuan Academy of Agricultural Sciences, and got PVP of the Ministry of Agriculture in 2017. It is now cultivated in Sichuan Province with a small amount of acreage.

Main Characters: Tree vigorous, natural spheroid, with strong budding and branching capacity. The fruiting twigs are mainly spring and early autumn shoots, prolific and yield stable. Fruit obovate, with a low collar; Rind orange, flesh dark orange, easy to peel. Fruit weight 236.8~359.8g, size in 8.2~9.4cm× 8.2~10.8cm, fruit shape index 0.95~1.23; Peel thickness 0.30~0.42cm, segments 12~14, on average 4.14 seeds per fruit in mixed cultivation; Edible portion 61.4%~71.0%, juicing rate 42.1%~57.2%, TSS 13.3%~16.2%, TA 0.4%~0.9%, Vitamin C 31.3~56.6mg/100ml. Pulp moderate in sweet and sour, tender with good mastication, flavor rich. Mature in Sichuan from the following February to March. Fruits are storage durable. Suitable areas for late-season citrus.

14. 金乐柑

来源与分布：金乐柑橘橙为不知火的优良变异，2007年在四川省金堂县发现。2021年获得植物新品种权。

主要性状：树势中庸，树姿较开张。萌芽力强，成枝力强，枝梢短粗。结果母枝以春梢和早秋梢为主，坐果率高，丰产稳产。果实倒卵形，有短果颈，果基有放射沟纹，果皮金黄色，果面较粗糙。单果重259.1~310.9g，大小为7.6~8.6cm×7.2~10.7cm，果形指数0.93~1.28，果皮厚度0.40cm，无核，可食率59.5%~76.4%，TSS11.0%~15.7%，酸0.8%~1.1%，每100ml果汁维生素C含量37.2~42.0mg，固酸比为12.2~18.7。果肉比不知火更加脆嫩，化渣，风味浓郁，品质优良。在四川4月中下旬成熟，比不知火晚1个月成熟，可留树保鲜至5月下旬。中抗溃疡病。适宜在冬季气候较冷凉、无冻害、土壤肥沃的川渝地区种植推广。

14. Jinlegan Tangor

Origin and Distribution: Jinlegan tangor is an elite mutation of Shiranui, found in Jintang County, Sichuan Province in 2007. The variety got PVP of the Ministry of Agriculture and Rural Affairs in 2021.

Main Characters: Tree medium in vigor, canopy somewhat open, budding capacity strong, branching capability strong, with short and strong twigs. Fruiting twigs are mainly spring and early autumn shoots. Fruit setting rate high, prolific and yield stable. Fruit obovate with low collar, base with radial furrows. Peel golden yellow, somewhat rough. Fruit weight 259.1~310.9g, size in 7.6~8.6cm×7.2~10.7cm, fruit shape index 0.93~1.28, peel thickness 0.40cm, seedless, edible portion 59.5%~76.4%, TSS 11.0%~15.7%, TA 0.8%~1.1%, Vitamin C 37.2~42.0mg/100ml, TSS/TA 12.2~18.7. Pulp crisper tenderer and of better mastication than that of Shiranui. Flavor rich, quality excellent. Maturity from mid-to-late April, one month later than that of Shiranui, and can be kept on trees until late-May. Moderately resistant to citrus canker. It is suitable for areas with fertile soil, a cool climate and frost-free winter such as Sichuan and Chongqing.

15. 金秋砂糖橘

来源与分布：金秋砂糖橘又名中柑所5号，是中国农业科学院柑橘研究所于2006年以红美人为母本、砂糖橘为父本杂交育成。2017年获得植物新品种权。

主要性状：树姿开张，枝梢节间短。春梢叶片大小为7.1cm×3.6cm，卵圆形，叶缘锯齿状，翼叶小或无。花中等大，完全花，花柱弯曲，花粉淡黄色。果实大小为4.6cm×3.8cm，单果重50～100g。果面橙红色，果皮薄而细腻，果皮韧性弱，易剥离。果肉橙色，细嫩化渣，风味纯甜。TSS约14.5%，酸0.3%，无核，自交不亲和，混栽有少量种子，成熟期10月下旬，是早熟品种，适宜在大部分柑橘产区种植。

15. Jinqiu Shatangju

Origin and Distribution: Jinqiu Shatangju, also known as CRIC No.5, was developed by crossing Beni Madonna (as the female parent) with Shatangju tangerine by Citrus Research Institute, Chinese Academy of Agricultural Sciences in 2006. This variety got PVP of the Ministry of Agriculture in 2017.

Main Characters: Tree spreading with short internodes. Leaf blade on spring shoot size in 7.1cm×3.6cm, ovate, margin dentate, petiole wing small or absent. Complete flower, medium-sized, style curved, with light-yellow pollen. Fruits size in 4.6cm×3.8cm, weight 50~100g, rind orange-red, thin and delicate, easy to peel. Pulp orange colored, tender and melting, taste pure very sweet with TSS 14.5%and TA 0.3%. Seedless, self-incompatible, a few seeds occur when mixed cultivated with other varieties. As an early-maturing cultivar, Jinqiushatangju matures in late October. The planting suitable areas include most areas for citrus production in China.

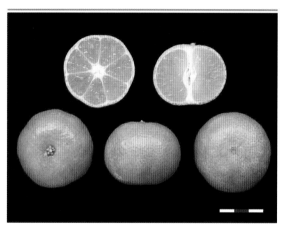

16. 凯旋柑

来源与分布：浙江省柑橘研究所用克里迈丁红橘为母本，与瓯柑和改良橙的杂种439橘橙（红柿柑）杂交育成。浙江省有少量栽培。

主要性状：树势中等，直立，结果后树势稍开张。春梢叶片大小为7.7cm×3.5cm，椭圆形。果实近圆球形，橙红或朱红色。TSS 16.9%～19.0%，酸1.1%，果肉鲜橙色、汁多，肉质脆嫩，味浓甜。种子8～12粒，成熟期11月上中旬。该品种丰产稳产，适合鲜食和加工果汁。

16. Kaixuangan

Origin and Distribution: Kaixuangan, originated from a cross between Clementine (female parent) and "439jucheng" (Ougan × Gailiangcheng), was selected by the Citrus Research Institute of Zhejiang Province, and is grown a little in Zhejiang Province.

Main Characters: Tree medium-vigorous, erect, slightly spreading after fruiting. Leaf 7.7cm×3.5cm, elliptic. Fruit nearly spheroid; orange-red or deep red. TSS 16.9%~19.0%, acid content 1.1%. Pulp bright orange, juicy, crisp, tender, rich sweet flavor. 8~12 seeds per fruit. Matures in early to mid-December. Yields high and consistent. It is good for both fresh consumption and juice processing.

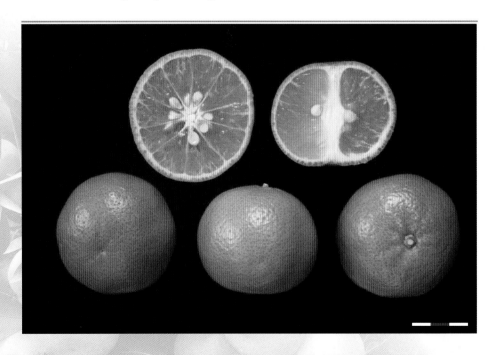

17. 明日见

来源与分布：原产日本，从（甜春橘柚×特洛维塔甜橙）×春见杂交后代中选育而成。主产四川、重庆、浙江等地。

主要性状：树势强，树姿开张，枝条具长刺。单果重180g，果实高扁圆形，果顶有放射沟；果皮橙色，较薄，略光滑，稍难剥离；果肉橙色，TSS14.0%～18.0%；肉质细嫩化渣，汁多味浓，可食率85%，出汁率55%，品质优良。少核或无核。1月下旬至2月下旬成熟，可延后到2～3月完熟（浙江地区）。对柑橘溃疡病敏感，植株抗寒性较强，耐-7℃低温。冬季低温在0℃以下地区完熟采收的，须采取保温大棚栽培。

17. Asumi

Origin and Distribution: Introduced from Japan, originated from the hybrid offspring of (sweet spring pomelo × Trovita sweet orange) × Harumi. Mainly grown in Sichuan, Chongqing, Zhejiang and other provinces.

Main Characters: Tree vigorous, spreading and thorny. Fruit weight 180g, long obloid, apex with radial furrows; pericarp orange, moderately thin and smooth, slightly difficult to peel; pulp orange, TSS 14.0%~18.0%; pulp juicy, tender and melting, edible portion 85% and juice rate 55%; few or no seeds. Maturity period ranges from late January to late February, and can extend for 2~3 months to fully mature in Zhejiang; Sensitive to citrus canker, strong cold tolerance to about −7℃. This variety is suitable for cultivation in the plastic-protected house or greenhouse in the area where the winter temperature is lower than zero.

18. 默科特

来源与分布：美国用橘和橙杂交育成。我国从美国、澳大利亚引入，台湾省当地称其为"茂谷柑"。福建、四川、湖北、广东等省有少量试种。

主要性状：树势旺盛，丛生分枝状树形，幼树稍直立，结果后树形逐渐开张。春梢叶片大小为6.7cm×3.5cm，椭圆形，叶缘锯齿不明显。花中大，完全花。果实高扁圆形，黄橙色，大小为6.9～7.0cm×5.2～5.3cm，皮薄，包着较紧，没有其他宽皮柑橘那样容易剥离，但皮较韧，也可剥离。TSS 12.0%，酸1.1%，肉质脆嫩，汁多，风味浓，品质优良。种子10～12粒/果，成熟期12月下旬至翌年2月上旬。该品种适应性广，早结丰产。

18. Murcott

Origin and Distribution: Murcott is a cross made between tangerine and orange in USA and was introduced to China from USA and Australia. It is called Maogugan in Taiwan Province. There are limited cultivations in Fujian, Sichuan, Hubei and Guangdong etc.

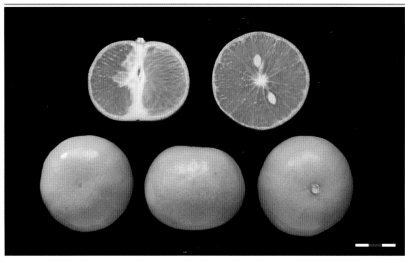

Main Characters: Tree vigorous, bush-like, young tree somewhat erect, gradually spreading after fruiting. Leaf 6.7cm×3.5cm, elliptic, margin dentate inconspicuous. Complete flower, medium sized. Fruit long obloid, yellowish orange, 6.9~7.0cm×5.2~5.3cm, rind thin, moderately adherent, more difficult than other mandarins to peel. TSS 12.0%, acid content 1.1%. Pulp crisp and tender, juicy, high in both sugar and acid content, rich flavor, excellent quality. 10~12 seeds per fruit. Maturity period ranges from late December to next February. The cultivar is widely adaptable, early fruiting and productive.

19. 南香

来源与分布：日本农林水产省于1970年用三保早生温州蜜柑与克里迈丁红橘杂交育成。我国20世纪末从日本引入，重庆、浙江等有少量试种。

主要性状：树势中等，直立，结果后开张。枝叶密集，春梢短且硬。春梢叶片大小为8.1cm×3.3cm，花中大，雄性不育，单性结果率高。果实高扁球形，橙红色，大小为150g左右，果顶凸起有小脐，果蒂部纵凸起有放射沟。果皮较薄，比温州蜜柑稍难剥离。TSS 13.0%～14.0%，酸1.0%。果肉柔软多汁，味浓，无核，成熟期12月中下旬。该品种适应性强，丰产性好，果实品质好，可以枳作砧木。

19. Nankou

Origin and Distribution: Nankou originated from a cross between Miho wase satsuma mandarin and Clementine, made by the Ministry of Agriculture, Forestry and Fisheries of Japan in 1970. It was introduced to China at the end of 20th century and is trial-grown in Chongqing and Zhejiang etc.

Main Characters: Tree medium-vigorous, erect, spreading after fruiting. Branches dense, spring shoots short and hard. Leaf 8.1cm×3.3cm, flower medium-large, male sterile, high rate of parthenocarpic fruit. Fruit long-obloid, orange-red, weight 150g; apex protruded, with small navel; base corrugated with radial grooves. Rind somewhat thin, slightly more difficult than Satsuma Mandarin to peel. TSS 13.0%~14.0%, acid content 1.0%; pulp tender, juicy, high sugar and high acid content, flavor rich, seedless. Matures in middle to late December. This variety is strongly adaptable and very productive. The fruit is of good quality. Common rootstock is Trifoliate orange.

20. 诺瓦

来源与分布：美国佛罗里达州用克里迈丁红橘与奥兰多橘柚杂交育成。我国于20世纪80年代从美国引入，四川、重庆、湖北等地有少量栽培。

主要性状：树势旺盛，明显呈宽皮柑橘特性，枝条多刺。叶片较狭，似柳叶。花中大，自交不亲和。果实扁圆形，橙红色，大小为5.4cm×5.0cm，果皮不易剥离。TSS 11.0%～12.0%，酸0.6%～0.7%，单一栽植时无核。果肉脆嫩，汁多、化渣、清甜，品质好。成熟期11月下旬至12月上旬。该品种果实耐贮性较差，枝条容易出现回枯现象。

20. Nova

Origin and Distribution: Nova originated from a cross between Clementine and Orlando tangelo made in Florida, USA. It was introduced to China from USA in 1980s and is cultivated limitedly in Sichuan, Chongqing and Hubei etc.

Main Characters: Tree vigorous, mandarin-like, branches thorny. Leaves somewhat narrow, like willow leaf. Flower medium-large, self-incompatible. Fruit obloid, orange-red, sizes in 5.4cm×5.0cm, difficult to peel. TSS 11.0%~12.0%, acid content 0.6%~0.7%, seedless. Pulp crisp, juicy, melting, tastes freshly sweet, of good quality. Maturity period ranges from late November to early December. Fruit stores poorly. Branchlets suffer easily from dieback.

21. 清见

来源与分布：日本农林水产省用特罗维塔甜橙（华脐实生变种）与宫川温州蜜柑（母本）杂交育成。我国20世纪80年代引进，四川、重庆、浙江、湖北、广东等地有少量试种。

主要性状：树势中等，幼树树姿稍直立，结果后逐渐开张。枝梢细长，易下垂。春梢叶片大小9.7cm×3.9cm，叶缘波状。花小，花枝大且弯曲。花药退化，花粉全无。果实扁球形，橙黄色，大小为8.2cm×7.1cm，果面较光滑。果皮稍难剥离，但仍可剥离。TSS 11.0%～12.0%，酸1.0%～1.3%，无核。果肉柔软多汁化渣，具甜橙香气，风味较佳。成熟期翌年3月上中旬。

21. Kiyomi

Origin and Distribution: Kiyomi originated from a cross between Miyagawa wase unshu (female parent) and Trovita sweet orange (a chance seedling of Washington navel) made by the Ministry of Agriculture, Forestry and Fisheries of Japan. In 1980s it was introduced to China and trial-cultivated in Sichuan, Chongqing, Zhejiang, Hubei and Guangdong etc.

Main Characters: Tree medium-vigorous, young tree relatively erect, becoming spreading after fruiting. Branches slender and long, tend to drop. Leaf 9.7cm×3.9cm, margin sinuate. Flower small, flower branches large and curved; anther degenerated without pollen. Fruit obloid, orange-yellow, 8.2cm×7.1cm, surface relatively smooth. Slightly difficult to peel. TSS 11.0%~12.0%, acid content 1.0%~1.3%, seedless. Pulp tender, juicy and melting with sweet-orange like aroma, good flavor. Matures in early to mid-March of next year.

22. 晴姬

来源与分布：日本农林水产省果树试验场现果树研究所兴津柑橘研究部以E647（清见×奥塞奥拉橘柚）为母本、宫川为父本杂交育成的柑橘品种。浙江省象山县于2002年从日本引入。

主要性状：在浙江省试验栽培过程中，该品种表现出糖度高、酸度低、囊衣薄、易剥皮等特点。幼树树姿较直立，结果后树冠开张，整体树势较为中等，枝条密且粗壮，节间较短，有短刺，结果后短刺逐渐退化。叶片较宫川、红美人小，平展或略上卷，纺锤形，单身复叶，边缘有钝或圆裂齿，叶色浓。果实扁圆形，单果重150~180g，果皮橙黄色，果肉橙色，果皮光滑，厚度约0.4cm，易剥离，不易浮皮，TSS12.0%~16.0%，囊衣薄，化渣性好，无核或少核，有橙子和橘子的甜味，口感柔和，有清香味。10月下旬开始转色，11月下旬完全着黄色，12月上旬转橙色，12月上中旬上市，设施栽培可完熟到翌年1月上中旬上市。坐果率较高，大小年结果现象不明显。适宜在冬季霜冻来临晚或利用设施栽培的长江以南温州蜜柑产区栽培。

22. Harehime Tangor

Origin and Distribution: Bred by the Okitsu Citrus Research Department of Fruit Tree Research Institute of the Ministry of Agriculture, Forestry and Fisheries of Japan, using E647 (Kiyomi × Osceola) as the female parent to cross with Miyagawa. Introduced to Xiangshan County, Zhejiang Province in 2002.

Main Characters: This variety showed the elite characteristics of high sugar content, low acid, thin membrane, easy to peel, etc. in the local cultivation trial. Tree medium vigorous, upright at youth, crown open after fruiting, branches dense and stout, with short internodes, thorns short and gradually absent after fruiting. Leaves smaller than Miyagawa and Beni Madonna, flat or slightly curling, fusiform, unifoliolate, margin bluntly or rounded dentate, color dark. Fruit obloid, average weight 150~180g; Rind orange-yellow, smooth, 0.4cm thick, easy to peel, no puffing; Flesh orange color, TSS about 12.0%~16.0%, segment membrane thin, melting, seedless or few seeds. Taste combined orange and mandarin, tender, with aroma. Color change starts in late October, completely yellow in late November, turning orange in early December, marketable in early-to-mid December. Fruits can hang on the tree until fully mature and marketable in mid-January of next year. Stable and productive with quite high fruit setting rate, alternate bearing non-distinctive. It is suitable for satsuma mandarin producing area where winter frost comes late or facility cultivation is used.

23. 秋辉

来源与分布：原产美国，由鲍尔橘柚与坦普乐橘橙杂交育成。我国1990年从美国引进。湖北、江西、重庆等地有少量栽培。

主要性状：树势中等，成枝力较强，枝梢纤细无刺易披垂。春梢叶片大小为8.9cm×3.4cm，披针形，果扁球形，深橙红色，大小为6.9～7.6cm×5.6～7.0cm，果顶带有一个内陷的小脐。果蒂部有放射沟纹，果面光滑，果皮薄易剥离。TSS 11.0%～12.0%，酸0.7%～0.9%，果肉柔软多汁，风味浓，品质佳。异花授粉后，种子30～40粒/果。成熟期10月底。成熟后留树易出现厚皮、枯水。

23. Fallglo

Origin and Distribution: Fallglo originated from a cross between Bower tangelo and Temple tangor in USA. It was introduced to China in 1990 and is limitedly cultivated in Hubei, Jiangxi and Chongqing etc.

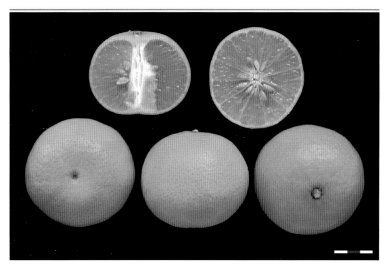

Main Characters: Tree medium-vigorous, strong in branching capability, branches slender, thornless and often drooping. Leaf 8.9cm×3.4cm, lanceolate. Fruit obloid, deep orange-red, 6.9~7.6cm×5.6~7.0cm, apex with a small depressed navel, base striped with radial grooves. Rind smooth, thin, easily peeling. TSS 11.0%~12.0%, acid content 0.7%~0.9%. Pulp soft, tender, juicy, rich flavor, quality excellent. 30~40 seeds per fruit when cross-pollinated. Matures in late October. Fruit will develop thick, puffy-skin and juice vesicle granulation if kept on tree after ripening.

24. 天草

来源与分布：日本农林水产省于1982年用清见×兴津早生14号的杂交后代为中间母本再与佩奇橘杂交育成。我国20世纪90年代从日本引入，四川、重庆、湖北、湖南、浙江、广东、广西等地有少量试种。

主要性状：树势中等，幼树稍直立，进入结果后树姿较开张。枝梢中等或偏密。春梢叶片大小为9.5cm×5.0cm，椭圆形。自花不育，能单性结果。果实扁球形，橙红色，表面光滑，有克里迈丁红橘和甜橙的香气，大小为7.1cm×5.3cm，果皮稍难剥离。TSS 11.0%～13.0%，酸0.7%～1.1%，品质优，风味好，无核。成熟期11月下旬至12月下旬。该品种早结丰产。以枳作砧，中间砧可用温州蜜柑、椪柑。

24. Amakusa

Origin and Distribution: Amakusa originated from a cross between Kiyomi tangor × Okisu wase 14 and Page tangor made by the Ministry of Agriculture, Forestry and Fisheries of Japan in 1982. China introduced Amakusa from Japan in 1990s. Amakusa is trial planted in Sichuan, Hubei, Hunan, Zhejiang, Guangdong and Guangxi etc.

Main Characters: Tree medium-vigorous, young tree somewhat erect, becoming spreading after fruiting, branches dense. Leaf 9.5 cm × 5.0 cm, elliptic. Infertile when self-pollinated, capable of parthenocarpy. Fruit obloid, orange-red, surface smooth, with aroma like Clementine and sweet orange, 7.1cm×5.3cm in size. Relatively difficult to peel. TSS 11.0%~13.0%, acid content 0.7%~1.1%. Pulp quality excellent, pleasant flavor and seedless. Matures from late November to late December. Early fruiting and productive. Commonly use Trifoliate orange as rootstock and Satsuma mandarin or Ponkan as interstocks.

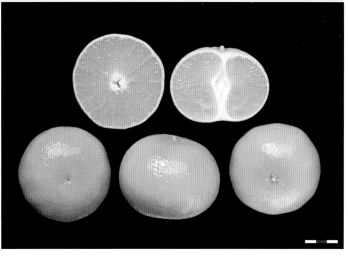

25. 沃柑

来源与分布：沃柑系从以色列引进的杂柑品种。20世纪70年代，以色列Vardi博士以坦普尔柑为母本与丹西红橘杂交而来。21世纪初引入我国，2010年后在广西南宁等地大面积推广。

主要性状：树势较旺，发枝力强。叶片浓密，浓绿色，阔披针形，叶片大小为8.4cm×3.5cm。果实扁圆形，单果重130g左右，大小6.7cm×5.7cm，果面光滑，橙色或橙红色，在干热河谷地区表现果皮油胞突出皮粗，果顶有不明显印圈。种子较多，10～20粒/果，果皮可以剥离，但较紧。TSS13.0%左右，高的可达16.0%或更高，酸0.5%～0.6%，每100ml果汁维生素C含量30～40mg。在广西南宁花期为3月底至4月初，采收期为翌年2～4月。该品种十分丰产，要求积温较高，冬季零度以下的气温概率低、有效积温≥6 000 ℃的地区较适宜。对疮痂病和褐腐病比较敏感。个别地方偶尔出现果皮白皮层变蓝现象，原因尚不明。

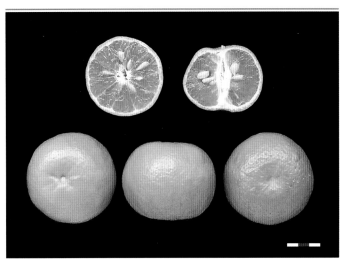

25. Orah

Origin and Distribution: Orah was bred via crossing Temple tangor as female with Dancy Red Tangerine (Dancy) by Dr. A.Vardi from Israel in the 1970s. Introduced into China at the beginning of this century and widely grown in Nanning and other places in Guangxi after the 2010.

Main characters: Tree quite vigorous, branching ability strong. Leaf dense, dark green, broadly lanceolate, size 8.4cm×3.5cm. Fruit obloid, weight 130g, size 6.7cm×5.7cm, rind smooth, orange or orange-red, apex with faint areole ring; rind turning rough with protruding oil glands in the dry-hot valley climate; Flavor rich, TSS 13.0% (sometimes 16.0% or even higher), TA 0.5%~0.6% and Vitamin C 40mg/100ml. Very prolific and seedy, 10~20 seeds per fruit. Rind adherent somewhat tight, peelable. Flowering from late March to early April in Nanning, Guangxi. Harvest from February to April in the following year. This variety requires a high-accumulated temperature. Generally, it can be grown in areas where the effective accumulated temperature is ⩾6 000 ℃. It is very sensitive to citrus scab and brown rot. Occasionally the albedo turns blue in some places for an unknown reason.

26. 阳光1号橘柚

来源与分布：阳光1号橘柚是中国农业科学院柑橘研究所以爱媛果试28号×春香杂交而来。2021年获得植物新品种权。

主要性状：树势强健，树姿开张，树冠圆头形。叶片阔披针形，叶缘有浅锯齿，翼叶倒披针形，叶基窄，叶面光滑，叶尖渐尖，少或无尖凹。完全花，花冠半张开，柱头弯曲，花药呈白色，无花粉，雄蕊完全败育。果实高扁圆形，果皮黄色，油胞细密，大部分果实果蒂部附近有明显的短颈，花柱容易宿存。中心柱半充实，皮厚度中等，包着紧实，果皮韧性极好，较易剥离。果肉橙色，囊瓣膜极薄，呈透明状，果肉脆嫩化渣，风味浓且酸甜可口。单果重255g，无核，TSS13.0%~15.0%，酸约为0.7%，每100ml果汁维生素C含量43.5mg，11月下旬成熟。回味甜，有轻微的柚类香气。果实田间较易开裂，但采果后极耐贮运。适宜在四川、重庆、广西、江西、云南、湖南、浙江种植。

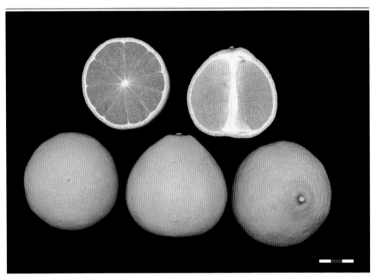

26. Yangguang No.1 Tangelo

Origin and Distribution: Yangguang No.1 Tangelo is a hybrid of Ehime Kashi No.28 × Haruka, the cross was made by Citrus Research Institute, Chinese Academy of Agricultural Sciences. The variety got PVP of the Ministry of Agriculture and Rural Affairs in 2021.

Main Characters: The plant is strong, crown is round and open posture. Leaf is broad-lanceolate, shallow serration in margin, oblanceolate wing leaves, narrow base, smooth surface, acuminate tip without or with little concave. The flower is a type of complete flower with half-open corolla, curved stigma, white anthers, no pollens and stamen aborted. The fruit is high oblate, with short neck near the pedicle, persistent style and semi-full central column. The peel is tough, tight and moderately easy to peel with dense oil glands, yellow and intermediate thickness. The flesh is orange, crisp, tender with very thin and transparent sac membranes, balanced sweet and sour taste, strong flavor with sweet aftertaste and slight aroma of pummelo. Fruit matures in late November. Fruit is prone to the cracking in the field, but good for storage and transportation. Some fruit quality indexes are as follows: seedless, fruit weight 255g, soluble solids 13.0%~15.0%, titratable acid 0.7%, Vitamin C 43.5mg/100ml. It is suitable to be planted in Sichuan, Chongqing, Guangxi, Jiangxi, Yunnan, Hunan and Zhejiang.

27. 伊予柑

来源与分布：原产日本，是橘柚类的自然杂种。我国20世纪70年代从日本引入，有3个品系，即宫内、大谷、胜山。浙江、重庆等地有少量试种。

主要性状：树势较强，树冠半圆头形，枝条粗壮，春梢叶片大小为8.4cm×4.1cm。果实倒卵形或高扁圆形，橙红色，大小为8.9cm×6.7cm。果面略显粗糙，大谷果面光滑，果皮较厚，易剥离。TSS 12.0%，酸1.0%，果肉柔软多汁，有芳香味，品质好。种子10～15粒/果。成熟期12月至翌年1月上旬。该品种适应性广，果实晚熟，可在树上挂果贮藏，采后也耐贮藏。以枸头橙、高橙、本地早等作砧木。

27. Iyokan

Origin and Distribution: Originated in Japan, Iyokan is a natural hybrid of tangerine and pummelo. Three lines were introduced from Japan to China in 1970s, including Miyauchi, Otani and Katsuyama. There are limited cultivations in Zhejiang and Chongqing.

Main Characters: Tree relatively vigorous, semi-spheroid, branches thick and strong. Leaf sizes in 8.4 cm×4.1cm. Fruit obovoid or long-obloid, orange-red, 8.9cm×6.7cm, surface somewhat rough, Otani surface smooth. Rind relatively thick, easily peeling. TSS 12.0%, acid content 1.0%. Pulp tender, juicy, fragrant aroma, of fine quality. 10~15 seeds per fruit. Maturity period ranges from December to early January. The variety is widely adaptable and late in maturity. Fruits store well both on tree and after harvested. Its common rootstocks are Goutoucheng, Gaocheng and Bendizao etc.

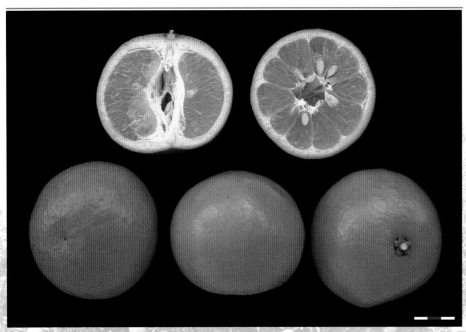

大谷伊予柑　Otani Iyokan

28. 媛小春

来源与分布：从日本引进，1994年日本爱媛县果树试验场以清见×黄金柑杂交而成，2007年申请日本品种登记。

主要性状：树势强健，初结果树较直立，挂果后自然圆头形、较开张，萌芽力中等，成枝力强。成花力弱，结果母枝以春梢或早秋梢顶部单花为主。果实圆球形，花期高温地区易导致果顶凸起呈高圆球形，果面略光滑，淡柠檬黄色，果皮易剥离。单果重125g，果肉柔软多汁，有特殊芳香味，无籽。1月下旬采收TSS13.0%左右，酸1.0%左右，可食率78.2%；完熟栽培到2月中下旬，TSS15.0%以上，酸0.8%以下，品质极佳。在浙江地区完熟栽培成熟期在翌年1月下旬至3月。丰产、稳产，与枳或温州蜜柑中间砧嫁接亲和性好。树势过强易导致投产推迟。适宜种植在冬季无明显霜冻危害的柑橘产区或用于设施栽培。

28. Himekoharu Tangor

Origin and Distribution: It was from the cross of Kiyomi × Ogokans in Ehime Prefecture, Japan in 1994, registered as a new variety introduced to China in 2007.

Main Characters: Tree vigorous, somewhat upright at the beginning of the fruiting, and gradually spreading after fruit setting, crown natural spheroid. Budding capacity moderate, branching capability strong. Flowering capacity weak, mainly spring or early autumn shoots bearing fruits. Fruit spherical, long spherical and apex protruded if flowering time meets high temperature. Seedless. Pericarp slightly smooth, light lemon yellow, easy to peel, fruit weight 125g, flesh juicy and tender, TSS 13.0%, TA 1.0%, edible portion 78.2%. If the fruits are delayed for harvest in mid-to-late February, TSS can reach above 15.0%, TA 0.8%, and has excellent quality. Over-vigorous trees would delay the fruit setting. Normally prolific and yield stable. Graft compatible with Trifoliate orange or using Satsuma as intermedia rootstock. Maturity from late January to March of next year in Zhejiang. It is suitable to be planted in the main producing areas where no obvious frost in winter or for protected cultivation.

29. 粤英甜橘

来源与分布：粤英甜橘是广东省农业科学院果树研究所从紫金县百年老橘园中选出的新株系，可能是柑与橘的自然杂种。在英德市推广种植较多。

主要性状：树势强，树冠呈半圆头形，枝条密集。春梢叶片大小为7.2～8.2cm×3.2～4.2cm，卵圆形。叶缘浅波状，翼叶较小，线形。花较小，完全花。果实高扁圆形，橙红色，大小为5.5～6.2cm×5.2～5.6cm，果顶平，微凹，果皮易剥离。TSS 13.8%，酸0.43%，果肉汁多，有蜜味。种子11～12粒/果。成熟期11月上旬，可在树上挂果贮藏到翌年1月上旬。该品种早结丰产性好，栽培上不用采取很多保果措施都能丰产。以酸橘作砧表现较好。

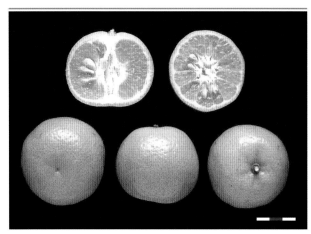

29. Yueyingtianju

Origin and Distribution: It is a new line selected from a hundred-year old citrus orchard in Zijin County by the Institute of Fruit Tree Research, Guangdong Academy of Agricultural Sciences. It is Probably a natural cross between Mandarin and Tangerine and planted mostly in Yingde of Guangdong.

Main Characters: Tree vigorous, semi-spheroid, branches dense. Leaf 7.2~8.2cm×3.2~4.2cm, ovate, margin shallowly sinuate; petiole wing relatively small, linear. Complete flower, smaller. Fruit long-obloid, orange-red, 5.5~6.2cm×5.2~5.6cm; apex truncate, slightly depressed; easily peeling. TSS 13.8%, acid content 0.43%. Pulp juicy with honey flavor. 11~12 seeds per fruit. Matures in early November. Fruits could hold on tree until early January. The variety is precocious and very productive. Its performance is well on Sour tangerine rootstock.

二、甜橙
Sweet Oranges

（一）普通甜橙 Common Sweet Orange

1. 奥林达夏橙

来源与分布：美国加利福尼亚州Webber氏在河边市柑橘研究中心从伏令夏橙的实生变异中选出的珠心系良种，我国于1979年从美国引进。现湖北、四川、重庆、广西、广东等地有少量栽培。

主要性状：果实椭圆形，橙红色，大小为6.8cm×7.1cm，较光滑。TSS 11.0%～12.0%，酸0.8%。种子4～6粒/果。成熟期翌年4月下旬至5月上旬。该品种较丰产，果实肉质脆嫩化渣，味甜有清香，品质较好。是鲜食加工兼用品种。

1. Olinda Valencia Orange

Origin and Distribution: The fine variety was selected from the nucellar seedlings of Valencia orange, by Webber in the Citrus Research Center, Riverside, California, USA, and was introduced to China in 1979. There are some cultivations in Hubei, Sichuan, Chongqing, Guangxi and Guangdong etc.

Main Characters: Fruit ellipsoid, orange-red, sizes in 6.8cm×7.1cm, rind comparatively smooth. TSS 11.0%~12.0%, acid content 0.8%. 4~6 seeds per fruit. Pulp crisp, tender and melting; taste sweet with fresh aroma. Maturity period ranges from late April to early May. The variety is relatively productive, and excellent for fresh consumption and juice processing.

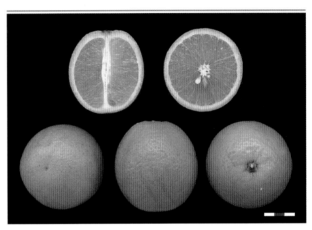

2. 冰糖橙

来源与分布：冰糖橙又名冰糖包，原产湖南黔阳，是当地普通甜橙芽变种。主产湖南省，四川、广东、广西、贵州等省（自治区）有少量栽培。

主要性状：树势中等，树冠半圆形。枝条细软无刺。春梢叶片大小为7.8～8.3cm×3.2～3.6cm，阔披针形。果实圆球形，橙黄色，大小为5.4～6.1cm×5.2～6.0cm，果顶圆钝，皮较难剥离。TSS 13.0%～15.0%，酸0.6%～0.8%。种子2～4粒/果。成熟期11月上中旬。果肉细嫩、汁多、化渣，味浓甜，品质优。1986年在普通冰糖橙中已选出麻阳大果冰糖橙和麻阳红皮大果冰糖橙，2006年均通过湖南省农作物品种审定委员会审定。

2. Bingtangcheng

Origin and Distribution: Bingtangcheng is also named as Bingtangbao and originated in Qianyang, Hunan Province, as a bud mutation of local common sweet orange. It is mainly grown in Hunan, and distributed in Sichuan, Guangdong, Guangxi and Guizhou etc.

Main Characters: Tree medium-vigorous, semi-spherical, delicate branches thornless. Leaf sizes in 7.8~8.3cm×3.2~3.6cm, broad lanceolate. Fruit spheroid, orange yellow, 5.4~6.1cm×5.2~6.0cm, apex obtusely rounded, moderately difficult to peel. TSS 13.0%~15.0%, acid 0.6%~0.8%, 2~4 seeds per fruit. Matures in early to middle November. Pulp tender, juicy, melting, with rich sweet flavor and excellent quality. The Mayang large-fruit Bingtangcheng and the Mayang red large-fruit Bingtangcheng, selected from common Bingtangcheng in 1986, were both registered as new varieties by Hunan Crop Registration Committee in 2006.

3. 长叶晚橙

来源与分布：长叶晚橙为长叶橙芽变，1998年在重庆市江津县支坪镇金沙寨长叶橙果园发现。2019年通过四川省农作物品种审定委员会审定。

主要性状：树势中庸，树冠圆头形；树姿直立，结果后逐渐开张，枝梢具有明显短刺。叶片披针形，叶形指数2.8，叶色浓绿。果实圆球形，单果重160g左右，果形整齐，果实大小为6.4～7.1cm×5.8～7.2cm，果形指数0.99；果皮橙黄色，果皮薄（0.38cm），包着紧，较难剥离。果肉橙黄色，汁胞细长，果肉细嫩，风味浓郁，有独特香气。单一品种连片种植时种子3～4粒/果。TSS11.1%～13.6%，酸0.4%～0.8%，固酸比15∶1至35∶1，每100ml果汁维生素C含量31.5～49.9mg，可食率75%。在重庆地区翌年2月中下旬果实成熟，果实挂树性能优，可挂树至6月上旬采收。加工性能优良，鲜食加工兼宜。该品种适宜在甜橙相似生态产区推广发展，尤其适宜在三峡库区作为晚熟鲜销与加工兼用品种推广。

3. Changye Wancheng

Origin and Distribution: Changye Wancheng orange is a bud mutation of Changye orange, found in Jiangjin County of Chongqing in 1998, and registered in Sichuan Province in 2019.

Main Characters: Tree medium in vigor, crown spheroid, upright and gradually spreading after bearing fruit, thorns present. Leaf lanceolate, dark green, shape index 2.8. Fruit spherical, weight 160g, shape uniform, size in 6.4~7.1cm ×5.8~7.2cm, fruit shape index 0.99; Rind orange-yellow, thin (0.38cm), tightly adherent, hard-to-peel. Flesh orange-yellow, juice sacs slender, pulp tender, flavor rich, aroma unique. Seeds 3 to 4 per fruit when the single variety is planted. TSS 11.1%~13.6%, TA 0.4%~0.8%, Vitamin C 31.5~49.9mg/100ml and edible portion 75%. In Chongqing, fruit matures in mid-to-late February in the following year and can be kept on trees to early June. Processing performance excellent, suitable for both fresh consumption and processing.

It is suitable for areas of navel oranges production especially the Three Gorges Reservoir area.

4. 长叶香橙

来源与分布：长叶香橙为普通长叶橙芽变优系选育而成，1994年在重庆市江津县白溪乡义安果园发现。2014年通过重庆市农作物品种审定委员会审定，2018年获得植物新品种权。

主要性状：树势较强，树冠自然圆头形。叶片披针形。果实圆球形，单果重170g左右，果形指数0.93～0.98。果面较光滑，油胞稀疏，果皮厚0.33～0.40cm。果肉细嫩化渣，甜味独特清爽，风味浓郁。果实无核或少核（0～4粒），多胚。TSS12.8%～13.4%，酸0.8%～1.0%，固酸比13～18，每100ml果汁维生素C含量43.8～54.8mg，可食率66.2%～73.5%，加工性能优良，鲜食加工兼宜。在重庆地区果实12月下旬成熟，果实耐贮藏，可留树保鲜到翌年4～5月。该品种适宜在甜橙相似生态产区推广发展，尤其适宜在三峡库区作为鲜销与加工兼用品种推广。

4. Changye Xiangcheng

Origin and Distribution: Changye Xiangcheng is a bud mutation of Changye orange, found in Jiangjin County, Chongqing in 1994, and registered in Chongqing and at the Ministry of Agriculture and Rural Affair in 2014 and 2018 respectively.

Main Characters: Trees quite vigorous, naturally spheroid. Leaf lanceolate. Fruit spherical, weight about 170 g, fruit shape index 0.93~0.98. Peel smooth, with sparse oil glands, 0.33~0.40 cm thick, pulp tender and melting, taste unique and refreshingly sweet, flavor rich, seedless or few (0~4 seeds), polyembryonic. TSS 12.8%~13.4%, TA 0.8%~1.0%, TSS/TA ratio (RTT) 13~18, Vitamin C 43.8~54.8mg/100ml, edible portion 66.2%~73.5%. Processing performance excellent, suitable for both fresh consumption and processing. Ripening in late December in Chongqing, fruit stores well and can be kept on trees until April to May the following year. Suitable production area: areas for sweet oranges production, especially the Three Gorges Reservoir area.

5. 大红甜橙

来源与分布：大红甜橙又名红皮甜橙，原产湖南黔阳，是普通甜橙的芽变种，主产湖南，四川、湖北、广东、广西等有分布。

主要性状：树较矮小，树冠圆头形，枝细软，刺少。春梢叶片大小为7.5～7.9cm×3.2～3.6cm，宽披针形。花较大，完全花。果实圆球形或椭圆形，深橙红色或大红色，大小为6.0～6.5cm×6.5～7.5cm，果顶圆，蒂部微凹。果皮较难剥离。TSS 11.0%～12.0%。果肉细嫩，汁多化渣，品质优良。种子5～10粒/果，成熟期11月中旬。该品种丰产性好，果实美观，耐贮运。已选出无核大红甜橙。

5. Dahong Sweet Orange

Origin and Distribution: Originated in Qianyang, Hunan Province. Dahong sweet orange is also named as Hongpi sweet orange. It is a bud sport of common sweet orange. Dahong sweet orange is mostly grown in Hunan and distributed in Sichuan, Hubei, Guangdong and Guangxi etc.

Main Characters: Tree comparatively stocky, spheroid, branches slender and tender, less thorny. Leaf sizes in 7.5~7.9cm×3.2~3.6cm, broad lanceolate. Complete flower, relatively large. Fruit spheroid to ellipsoid, deep red-orange or bright red, 6.0~6.5cm×6.5~7.5cm in size, apex rounded, base slightly depressed. Comparatively difficult to peel. TSS 11.0%~12.0%. Pulp is tender, juicy and melting. 5~10 seeds per fruit. Matures in mid-November. The variety is quiet productive. Fruit beautiful, stores and ships well. Seedless variation has been selected.

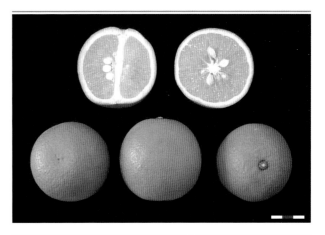

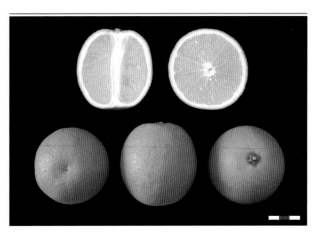

无核大红甜橙　Seedless Dahong sweet orange

6. 德塔夏橙

来源与分布：南非从伏令夏橙实生变异中选出。我国20世纪末从美国引进，现湖北、重庆、四川等地有少量栽培。

主要性状：果实略大于伏令夏橙，椭圆形，果皮光滑，橙红色，大小为6.5cm×6.8cm。TSS 11.5%，酸9.5%，无核。成熟期比伏令夏橙早1～3周。该品种以内膛挂果为主。

6. Delta Valencia Orange

Origin and Distribution: It was selected from natural mutations of Valencia orange in South Africa. China introduced it from America at the end of 20th century and is currently planted in Hubei, Chongqing and Sichuan etc.

Main Characters: Fruit slightly larger than Valencia, ellipsoid, rind smooth, orange-red, sizes in 6.5cm×6.8cm, TSS 11.5%, acid 9.5%, seedless. It matures 1~3 weeks earlier than Valencia. The tree has a tendence of bearing fruit in the inside of the canopy.

7. 伏令夏橙

来源与分布：伏令夏橙又名晚生橙、晚令夏橙，原是我国甜橙（很可能是雪柑）的变异种。据美国遗传杂志刊载，15世纪中叶传入欧洲西班牙的瓦伦西亚等地栽培，19世纪70年代再由英国苗商从亚速尔岛输入美国后得到大量发展，成为美国和世界柑橘类的重要品种。我国于20世纪40年代从美国及摩洛哥等国引进。现主要在四川、广东、广西、福建、湖北、云南等地有少量栽培。

主要性状：树势中等，树冠圆头形或塔形，枝条粗壮，无刺或少刺。春梢叶片大小为9.0～9.5cm×3.8～4.0cm，长卵圆或长椭圆形，叶缘无锯齿，翼叶小。花中大，花粉部分败育或胚囊不育。果实椭圆形或圆球形，单果重150～250g，橙黄色或橙红色。TSS 11.0%～12.0%，酸1.2%～1.3%。种子0～6粒/果。成熟期翌年4月下旬至5月上旬。该品种丰产性中等，是加工橙汁最好的原料。主产国已从其芽变选出不少新品系。

7. Valencia Orange

Origin and Distribution: Valencia orange is also named as Wansheng orange, Wanglingxia orange. According to American genetic magazine, Valencia orange originated as a mutation of Sweet orange of China (most likely Xuegan sweet orange), and was introduced to Valencia, Spain in the middle of the 15th century. In 1870s, it was imported to USA from Azores Island by British nurserymen and has achieved great development there, becoming gradually a very important orange variety in America and the world. China introduced the variety from America and Morocco in 1940s. It is now cultivated limitedly in Sichuan, Guangdong, Fujian, Hubei and Yunnan.

Main Characters: Tree medium-vigorous, spheroid to tower-like, branches stout, thornless or less thorny. Leaf 9.0~9.5cm×3.8~4.0cm, long-ovate or long-elliptical, margin edentate, petiole wing small. Flower medium-sized, pollen and megaspore partially sterile. Fruit ellipsoid or spheroid, weight 150~250g. Color orange-yellow to orange-red. TSS 11.0%~12.0%, acid content 1.2%~1.3%, 0~6 seeds per fruit. Maturity period ranges from late April to early May. It is medium-productive and one of the best varieties for juicy processing. Many new lines have been selected from bud mutations in the major Valencia orange producing countries.

8. 福罗斯特夏橙

来源与分布：美国加利福尼亚州柑橘研究中心从伏令夏橙的实生苗中选出的珠心变异良种。我国1965年从摩洛哥引进。现四川、重庆、广西、福建、浙江、江西、湖北、湖南等地有试种。

主要性状：果实圆球形或椭圆形，橙黄色，大小为6.1cm×6.4cm。TSS 12.0%，酸1.1%。种子0～2粒/果。成熟期4～5月。该品种风味浓，品质优，是鲜食和加工兼用品种。

8. Frost Valencia Orange

Origin and Distribution: The fine cultivar was selected from a nucellar seedling of Valencia orange by the Citrus Research Center of California, USA. China introduced it from Morocco in 1965. Currently it is trial-planted in Sichuan, Chongqing, Guangdong, Guangxi, Fujian, Zhejiang, Jiangxi, Hubei and Hunan etc.

Main Characters: Fruit spheroid to ellipsoid, orange-yellow, sizes in 6.1cm×6.4cm, TSS 12.0%, acid content 1.1%, 0~2 seeds per fruit. Maturity period ranges from April to May. The variety is rich in flavor, of excellent quality and suitable for fresh consumption and juicy processing.

9. 改良橙

来源与分布：改良橙又名漳州橙，原产福建漳州。属印子柑和红橘的嫁接嵌合体，主产福建等省。

主要性状：树势强健，树冠圆头形，枝细而密。春梢叶片大小为8.1～9.5cm×3.7～4.8cm，长椭圆形。花为完全花。果实圆球形，橙黄色，大小为6.1～6.4cm×5.7～6.4cm，果顶平，有明显的印圈。果实有红肉型、黄肉型、黄肉或红肉相间3种类型，个别还会出现红橘果。果皮难剥离。TSS 12.0%～14.0%，酸0.9%～1.1%，种子15～18粒/果。成熟期12月中下旬。红肉果的肉质细嫩，化渣，多汁，酸甜适口，深受消费者欢迎。该品种适应性强，主要以福橘作砧木。

20世纪60年代，广东廉江县国营红江农场从广西引进甜橙苗种植后发现有改良橙并从中选出，80年代定名为红江橙，主产广东等省。以酸橘作砧。广东省农业科学院果树研究所以及华南农业大学园艺系选出无核或少核红江橙。

9. Gailiang Orange

Origin and Distribution: Gailiang orange, named also as Zhangzhoucheng, originated in Zhangzhou, Fujian Province. It is a graft chimera of Yinzigan and Red orange. It is mainly grown in Fujian.

Main Characters: Tree vigorous, spheroid, branches slender and dense. Leaf 8.1~9.5cm× 3.7~4.8cm in size, long-elliptical. Complete flower. Fruit spheroid, orange-yellow, sizes in 6.1~6.4cm× 5.7~6.4 cm, apex flattened with a prominent areole ring. Three types of pulp colors: red, yellow, and red-yellow chimera. Occasionally bears mandarin fruit. Difficult to peel. TSS 12.0%~14.0%. acid content 0.9%~1.1%, 15~18 seeds per fruit. Matures in middle to

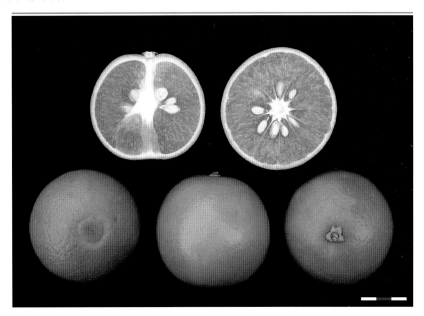

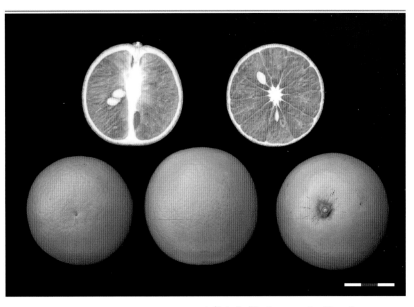

少核红江橙　Less seedy Hongjiang orange

late December. The variety is widely adaptable. Fruit with red pulp tender and melting with well blended sour-sweet flavor, and well accepted by consumers. Fuju is its common rootstock.

A better line was selected in 1960s by Hongjiang State Farm, in Lianjiang, Guangdong Province, from Gailiang orange seedlings introduced from Guangxi. This selection was named "Hongjiang orange" in 1980s. Hongjiang orange is mainly grown in Guangdong Province. Sour tangerine makes a good rootstock.

Seedless and less seedy Hongjiang orange lines have been selected by the Institute of Fruit Tree Research, Guangdong Academy of Agricultural Sciences and by the Department of Horticulture, Huanan Agricultural University.

10. 桂橙1号

来源与分布：桂橙1号是冰糖橙优良芽变株系，于广西鹿寨县四排乡甜橙果园发现。2008年通过广西农作物品种审定委员会审定。

主要性状：树势中等，树形开张。结果母枝以秋梢为主，占70%以上。单果重154.0g，果形指数0.98，果实横径6.0cm以上占36.1%，5.5～6.0cm占53.3%，5.5cm以下占10.6%。果皮厚0.41cm，种子1.3粒/果。果汁率50.0%，可食率75.7%，每100ml果汁维生素C含量46.0mg，柠檬酸0.4%，全糖12.5%，TSS14.6%。风味浓郁，有蜜香，甜脆化渣，果实耐贮藏。丰产稳产、适应性强，在鹿寨县成熟期11月下旬。适宜在甜橙种植区推广。

10. Guicheng No.1

Origin and Distribution: Originated from a bud mutation of Bingtangcheng, found in Sipai Town, Luzhai County, Guangxi; it was registered in Guangxi in 2008.

Main Characters: Trees medium in vigor, spreading. Fruiting twigs are mainly autumn shoots. Fruit weight 154.0g, fruit shape index 0.98; 36.1% of fruits with a transverse diameter of above 6.0cm, 53.3% with 5.5 to 6.0cm, and 10.6% with below 5.5cm; Rind thickness 0.41cm, seeds 1.3 per fruit, juicing rate 50.0%, edible portion 75.7%. TSS 14.6%, total sugar content 12.5%, TA 0.4%, Vitamin C 46.0mg/100ml. Flavor rich, of honey aroma, sweet, crisp, melting, storage durable. Prolific and yield stable, with good adaptability. Maturity in late November in Luzhai, Guangxi. Suitable for sweet orange cultivation area.

11. 桂夏橙1号

来源与分布：桂夏橙1号是从阿尔及利亚夏橙芽变中选育的新品种，发现于广西灵川县大境乡，2015年通过广西农作物品种审定委员会审定。

主要性状：果实圆球形或椭圆形，果皮橙黄色，果面较粗糙，果顶圆。果肉橙黄色，味酸甜，多汁，有香气。果大，单果重197.8～259.7g，大小为7.1～8.0cm×6.9～8.0cm，果形指数0.94～1.03；种子1.7粒/果，一般1～3粒。可食率70.3%～76.5%，果汁率45.0%～54.4%，TSS10.2%～13.3%，酸0.7%～1.2%，全糖7.2%～9.7%，固酸比9.9～15.1，每100ml果汁维生素C含量46.0～55.0mg。果实较化渣，品质好。果实生育期320～380d，在桂林灵川3月下旬至5月中下旬成熟。适宜在广西桂林以南无霜冻的地区栽培。

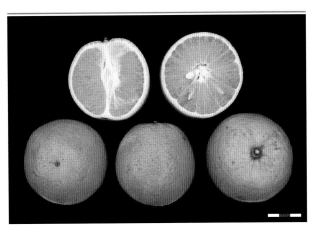

11. Guixiacheng No.1 Summer Orange

Origin and Distribution: Guixiacheng No.1 summer orange was selected from a bud mutation of Valencia orange introduced from Algeria, found in Dajing, Lingchuan County of Guangxi and registered in Guangxi in 2015.

Main Characters: Fruit spheroid or ellipsoid, rind orange-yellow, surface rough. Apex round, pulp orange-yellow, flavor sour and sweet, juicy, fragrant. Fruit big, weight 197.8~259.7g, size in 7.1~8.0×6.9~8.0 cm, fruit shape index 0.94~1.03; Seeds 1~3 per fruit. Edible portion 70.3%~76.5%, juicing rate 45.0%~54.4%, TSS 9.9%~15.1%, TA 0.7%~1.2%, total sugar 7.2%~9.7%, TSS/TA 9.9~15.1, Vitamin C 46.0~55.0mg/100ml. Maturity in late March to late May in Lingchuan, Guilin. Suitable for frost-free areas in south of Guilin, Guangxi.

12. 哈姆林甜橙

来源与分布： 原产美国，1965年从摩洛哥引入我国。广东、广西、四川、重庆、浙江、福建等地有少量栽培。

主要性状： 生长势中等，树冠半圆形，枝条密生，有短刺。春梢叶片大小为8.2～9.0cm×3.8～4.2cm，长椭圆形，叶缘波状，翼叶小。花中大，完全花。果实圆球形，橙黄色，极光滑。TSS 11.0%～12.0%，酸0.6%～0.8%。种子3～4粒/果，果肉汁多，较化渣。成熟期10月下旬至11月上旬。该品种早结丰产性好，是加工果汁和鲜食兼用的优良早熟品种。

12. Hamlin Sweet Orange

Origin and Distribution: Originated in USA, it was introduced to China from Morocco in 1965. There are some cultivations in provinces of Guangdong, Guangxi, Sichuan, Chongqing, Zhejiang and Fujian etc.

Main Characters: Tree medium-vigorous, semi-spheroid, branches dense, with short spines. Leaf blade 8.2~9.0cm×3.8~4.2 cm in size, long elliptical, margin sinuate, petiole wing small. Complete flower, medium-large. Fruit spheroid, color orange-yellow, surface rather smooth. TSS 11.0%~12.0%, acid 0.6%~0.8%, 3~4 seeds per fruit, pulp juicy and comparatively melting. Maturity period ranges from late October to early November. The variety is early fruiting, quite productive, and good for fresh consumption and juice processing.

13. 化州橙

来源与分布：原产广东化州，是20世纪70年代化州主栽品种之一。

主要性状：树冠呈不规则圆头形，枝梢粗壮，常具短刺。春梢叶片大小为8.5～9.5cm×4.5～5.2cm，呈卵状长椭圆形，翼叶不明显。花中大，完全花。果实圆球形或高扁圆形，橙黄色，大小为5.8～6.5cm×5.6～6.2cm，果皮难剥离。TSS 10.0%～13.0%，酸0.6%～0.9%。种子8～15粒/果。多汁化渣，甜酸适中，品质中。该品种早结丰产性好，是鲜食和加工制汁兼用品种。

13. Huazhou Orange

Origin and Distribution: Huazhou orange originated in Huazhou, Guangdong Province and was one of the most widely grown varieties in Huazhou in 1970s.

Main Characters: Tree irregularly spheroid, branches stout, frequently with short spines. Leaf 8.5~9.5cm×4.5~5.2cm in size, ovate to long-elliptical, petiole wing inconspicuous. Complete flower, medium-large. Fruit globose to long oblate, orange-yellow, sizes in 5.8~6.5cm×5.6~6.2cm, difficult to peel. TSS 10.0%~13.0%, acid content 0.6%~0.9%, 8~15 seeds per fruit. Flesh is juicy, melting, moderate sweet sour and of medium quality. The variety is early fruiting, quite productive, and good for both fresh consumption and juice processing.

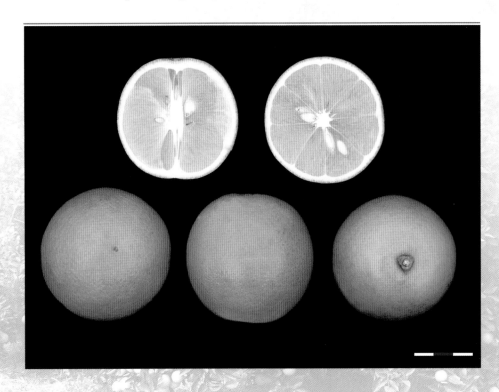

14. 锦橙

来源与分布：锦橙又名鹅蛋柑26号，原产重庆江津。是20世纪40年代从地方实生甜橙中选出的优良变异。主产重庆、四川、湖北。

主要性状：树势强健，树冠圆头形，树姿开张，枝有短刺。春梢叶片大小中等，长椭圆形。花中大，完全花。果实长椭圆形，形似鹅蛋，橙黄色或深橙色，大小为5.0～5.5cm×5.9～6.3cm，果顶平或微凹，果皮难剥离。TSS 11.0%～11.5%，酸0.7%～1.1%，种子6粒/果。成熟期11月下旬至12月上旬。该品种丰产性中等，品质优。果实耐贮藏。砧木用枳或红橘。

14. Jincheng

Origin and Distribution: Jincheng is also named as Edangan No.26 and originated in Jiangjin, Chongqing. This fine variety was a chance seedling selected in 1940s from local common sweet orange. It is now mainly grown in Chongqing, Sichuan and Hubei.

Main Characters: Tree vigorous, spheroid, spreading, with short spines on branches. Leaf medium-sized, long elliptic. Complete flower, medium-sized. Fruit long-elliptical, goose-egg like, orange yellow or deep orange, 5.0~5.5cm×5.9~6.3cm, apex flattened or slightly depressed, difficult to peel. TSS 11.0%~11.5%, acid 0.7%~1.1%, 6 seeds per fruit. Maturity period ranges from late November to early December. Trees are moderately productive; fruit quality fine; stores well. Trifoliate orange and Hongju are its common rootstocks.

15. 锦红冰糖橙

来源与分布：锦红是从普通冰糖橙芽变中选育出的冰糖橙新品种，1986年在湖南省麻阳县车头村果园发现，2007年通过湖南省农作物品种审定委员会审定。

主要性状：树势较强，树冠自然圆头形。枝条具短刺，萌芽力强，抽梢量大。结果母枝以春梢或早秋梢为主。果实圆形或近球形，少核或无核，果面橙红色，果肉浅橙色，单果重150～210g，TSS14.0%，酸0.6%，每100ml果汁维生素C含量51mg。锦红与枳、温州蜜柑、椪柑等嫁接亲和，丰产、稳产，在湖南省麻阳县12月中下旬完全成熟。适宜年均温达到16.5℃、年积温达到5 200℃以及最低温高于-5℃的全国甜橙优势栽培区和其他甜橙种植区。

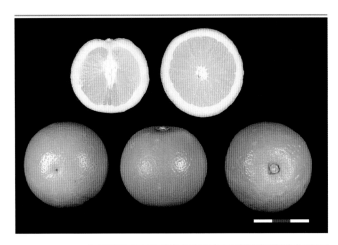

15. Jinhong Bingtangcheng

Origin and Distribution: Jinhong was selected from a bud mutation of common Bingtangcheng, found in an orchard in Chetou village, Mayang County, Hunan Province in 1986, and registered in 2007.

Main Characters: Strong tree vigor, crown natural spheroid, short spines, strong budding capacity. The fruiting twigs are mainly spring shoots or early autumn shoots. Fruit round or nearly spherical, seedless or few seeds, peel orange-red, pulp light orange; fruit weight 150~210g, with TSS 14.0%, TA 0.6% and Vitamin C 51mg/100ml. It is graft-compatible to Trifoliate Orange, Satsuma Mandarin and Ponkan. Stably prolific. Fully matures in the middle to late December in Mayang County, Hunan Province. Suitable for the area where sweet oranges can be grown, with an average annual temperature above 16.5℃, annual accumulated temperature above 5 200℃, and minimum temperature above -5℃.

16. 锦秀冰糖橙

来源与分布：锦秀冰糖橙是从普通冰糖橙芽变中选育出的新品种。2007年在湖南省麻阳县舒家村乡桐坡村果园发现。2016年获得植物新品种权。

主要性状：树势强，树姿开张，成枝力强。成花能力强，结果母枝以春梢或早秋梢为主。果实扁圆形，果面光滑，果实少核或无核，果面橙色，单果重188g，果肉浅橙色，柔软多汁，TSS12.0%以上，酸0.2%左右，每100ml果汁维生素C含量58mg，可食率为75%。锦秀冰糖橙与枳、温州蜜柑嫁接亲和，丰产、稳产，在湖南省麻阳县11月下旬成熟。适宜年均温达到16.5℃、年积温达到5 200℃以及最低温高于-5℃的全国甜橙优势栽培区。

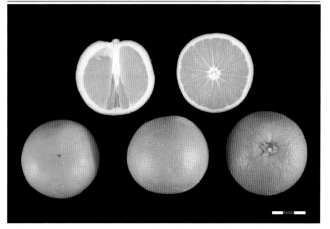

16. Jinxiu Bingtangcheng

Origin and Distribution: Jinxiu was selected from a bud mutation of Bingtangcheng, and found in an orchard in Tongpo, Shujiacun, Mayang County, Hunan Province in 2007. This variety got PVP of the Ministry of Agriculture in 2016.

Main Characters: Tree vigorous, spreading, with strong branching capacity. Flowering ability is strong, and the fruit twigs are mainly spring or early autumn shoots. Fruit obloid, seedless or few seeds; peel smooth, orange-red, fruit weight 188g on average. Pulp light orange color, tender and juicy, with a TSS over 12.0%, TA 0.2%, Vitamin C 58mg/100ml and edible portion 75%. This variety is graft-compatible to trifoliate orange, and Satsuma; prolific and yields stable. Maturity in Mayang County of Hunan Province in late November. It is suitable for where the sweet orange can grow, with an average annual temperature above 16.5℃, an annual accumulated temperature of 5 200℃, and minimum temperature higher than -5℃.

17. 橘湘珑冰糖橙

来源与分布： 橘湘珑冰糖橙是1995年从冰糖橙胚芽嫁接苗中选育的新品种。2018年获得植物新品种权。

主要性状： 树势强，树姿开张，成枝力强，枝梢密度很大。成花力强，结果母枝以秋梢为主。果实近球形，大小均匀，无核或种子数极少，果面橙色，单果重120～150g，果肉浅橙色，细嫩化渣，TSS14.5%，酸0.4%，每100ml果汁维生素C含量47.0mg，可食率为70%。橘湘珑冰糖橙与枳、温州蜜柑及冰糖橙均嫁接亲和，丰产、稳产，在湘南地区11月下旬完全成熟。适宜年均温达到16.5℃、年积温达到5 200℃以及最低温高于-5℃的甜橙种植区。

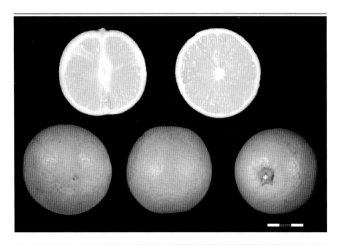

17. Juxianglong Bingtangcheng

Origin and Distribution: Selected from the nucellar seedlings of Qianyang Bingtangcheng grafted on trifoliate rootstock in 1995 and got PVP of the Ministry of Agriculture and Rural Affairs in 2018.

Main Characters: Tree vigorous and spreading, with strong branching capacity and dense canopy. Flowering ability is strong, and the fruit twigs are mainly autumn shoots. Fruit uniform in size and nearly spherical, seedless or few seeds, orange peel, average weight 120~150g. Pulp light orange, tender and melting, with TSS 14.5%, TA 0.4%, Vitamin C 47.0mg/100ml, and edible portion 70%. Graft-compatible with the Trifoliate Orange, Satsuma Mandarin and Bingtangcheng sweet orange. Stably prolific and fully matures in late November in southern Hunan. It is suitable for sweet orange production area where the average annual temperature is above 16.5℃, annual accumulated temperature over 5 200℃, and minimum temperature higher than -5℃.

18. 橘湘元糖橙

来源与分布：橘湘元糖橙是从埃及糖橙芽变中选育的无核糖橙新品种，2007年在湖南农业大学教学实习基地发现。2019年获得植物新品种权。

主要性状：树势强健，树冠自然圆头形，较开张，成枝力强。成花力强，结果母枝以春梢或早秋梢为主。果实圆球形，果形指数为0.95。单果重173g，大小较均匀，无核或少核，果面较光滑，浅黄色，果肉橙黄色。TSS13.6%，酸0.1%，每100ml果汁维生素C含量60.6mg，可食率达77.1%，果实出汁率54.6%。橘湘元糖橙与枳、温州蜜柑及其他甜橙均嫁接亲和，丰产、稳产，在湘南地区11月下旬完全成熟。适宜年均温达到16.5℃、年积温达到5 200℃以及最低温高于-5℃的甜橙优势栽培区和其他甜橙种植区。

18. Juxiangyuan Sweet Orange

Origin and Distribution: A seedless bud mutation of Succari Sweet Orange, was found in 2007 in an orchard at Hunan Agricultural University, China. This variety got PVP of the Ministry of Agriculture and Rural Affairs in 2019.

Main Characters: Tree vigorous, crown naturally rounded, canopy somewhat open, branching capacity strong. Flowering ability is strong, and fruit twigs are mainly spring or early autumn shoots. Fruit spherical, fruit shape index 0.95; Average fruit weight 173g, size uniform, seedless or few seeds. Surface somewhat smooth, rind light-yellow, flesh orange-yellow; TSS 13.6%, TA 0.1%, Vitamin C 60.6mg/100ml, edible portion 77.1%, and juicing rate 54.6%. Graft compatible with Trifoliate Orange, Satsuma and other sweet oranges. Prolific and yield stable. Full maturity in late November in southern Hunan. It is suitable for sweet orange producing area with an average annual temperature above 16.5℃, an annual accumulated temperature of 5 200℃, and the minimum temperature higher than -5℃.

19. 卡特夏橙

来源与分布：1935年Fawcett在美国加利福尼亚州河边市柑橘研究中心从实生苗变异中选出的珠心变异优良品种，我国1977年从美国引进。现湖北、重庆、四川等地有少量栽培。

主要性状：果实球形，橙黄色，大小为6.6cm×6.8cm，果面稍粗糙。TSS 10.0%，酸1.0%以上。种子9粒/果。成熟期4月下旬至5月上旬。该品种丰产性好，品质上等。

19. Cutter Valencia Orange

Origin and Distribution: The fine variety was selected from nucellar seedlings by Fawcett in the Citrus Research Center, Riverside, California in America in 1935. China introduced it from America in 1977 and has some plantings in Hubei, Chongqing and Sichuan etc.

Main Characters: Fruit spheroid, orange-yellow, sizes in 6.6cm×6.8cm, surface somewhat rough. TSS 10.0%, acid content above 1.0%, 9 seeds per fruit. Maturity period ranges from late April to early May. The variety is quite productive and of superior quality.

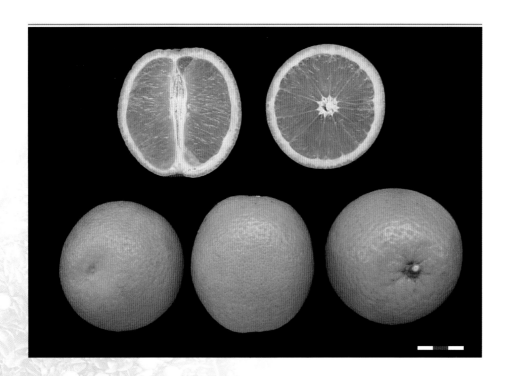

20. 柳橙

来源与分布：柳橙又名雷橙（广东）、印子柑（福建）、金印橙（台湾）。原产广东新会，经长期栽培筛选，有暗柳橙、明柳橙、半柳橙。主产广东、广西、福建和台湾，四川、浙江等省也有引种。

主要性状：长势强健，树冠半圆形，较开张。枝梢紧密，健壮。春梢叶片大小为7.9～10.9cm×3.0～4.4cm，呈长椭圆或椭圆形，叶缘全缘多呈波状。花中等大，完全花。果实圆球形或长圆形，橙黄色或橙红色，大小为5.7～6.5cm×5.5～6.0cm，果顶圆，印圈大而不明显，果皮不易剥离。TSS 12.0%～15.0%，酸0.6%～0.7%。种子5～15粒/果。果肉脆，较化渣，味清甜，汁中等，有香气。成熟期11月下旬至12月上旬。该品种丰产稳产性好，适应性广。主要砧木为酸橘、三湖红橘。

20. Liucheng

Origin and Distribution: Liucheng is also named as Liucheng (Guangdong), Yinzigan (Fujian), Jinyincheng (Taiwan) and originated in Xinhui, Guangdong. Many lines have been selected during long-term cultivation, including Anliucheng, Mingliucheng, and Banliucheng. The variety is mainly grown in Guangdong, Guangxi, Fujian, Taiwan, and also introduced to Sichuan and Zhejiang.

Main Characters: Tree vigorous, semi-spherical, relatively spreading, branches stout and thick. Leaf 7.9~10.9cm×3.0~4.4cm in size, long-elliptical to elliptical, margin entire and mostly sinuate. Complete flower, medium sized. Fruit spheroid to long- spheroid, orange yellow or orange red, 5.7~6.5cm×5.5~6.0cm. Apex rounded, areole ring large and faint. Difficult to peel. TSS 12.0%～15.0%, acid 0.6%~0.7%, 5~15 seeds per fruit. Pulp crisp, relatively melting, flavor fresh sweet, moderately juicy, fragrant. Maturity period ranges from late November to early December. The variety is consistently productive and widely adaptable. Its common rootstocks are Sour tangerine and Sanhuhongju.

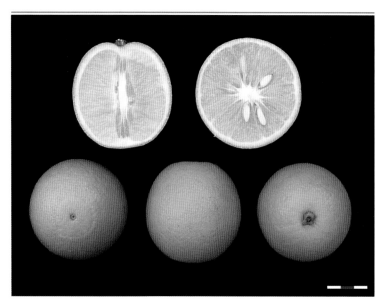

（1）暗柳橙　果实圆球形或椭圆形，橙红色，大小为5.6～6.5cm×5.5～6.0cm，果顶多有印圈。TSS 12.0%～14.0%，酸0.4%～0.7%。种子10～13粒/果。较化渣，味清甜，有香气。成熟期11月下旬至12月上旬。该品种适应性强，早结丰产，品质中上。

（1）Anliucheng　Fruit spheroid to elliptical, red-orange, sizes in 5.6～6.5cm×5.5～6.0cm, apex mostly with areole ring. TSS 12.0%～14.0%, acid content 0.4%～0.7%, 10～13 seeds per fruit. Relatively melting, flavor fresh sweet, fragrant. Maturity period ranges from late November to early December. The variety is adaptable, productive and early in maturity, quality medium or higher.

（2）**丰彩暗柳橙** 丰彩暗柳橙是广东省农业科学院果树研究所和杨村柑橘场从暗柳甜橙的实生后代中选出，1984年通过广东省农作物品种审定委员会审定。在广东、广西、福建等省（自治区）有栽培。

果实圆形或近圆形。橙红色，大小为5.5～6.4cm×5.4～6.1cm，果顶有印圈。TSS 12.3%～13.0%，酸0.8%～0.9%。种子13～15粒/果。汁多，风味浓郁，成熟期12月中旬。该品种丰产稳产，适应性强，果实含高糖高酸，较耐贮藏。

(2) Fengcai Anliucheng Fengcai Anliucheng was selected as a chance seedling from Anliu sweet orange by the Institute of Fruit Tree Research, Guangdong Academy of Agricultural Sciences and Yangcun Citrus Farm. It was registered by Guangdong Crop Variety Registration Committee in 1984. There are some cultivations in Guangdong, Guangxi, and Fujian etc.

Fruit nearly spheroid to spheroid, red-orange, sizes in 5.5~6.4cm× 5.4~6.1cm, apex marked with areole ring. TSS 12.3%~13.0%, acid content 0.8%~0.9%, 13~15 seeds per fruit. Juicy and rich flavor. Matures in mid December. The variety is consistently productive, adaptable. Fruit is high in both sugar and acid content and stores relatively well.

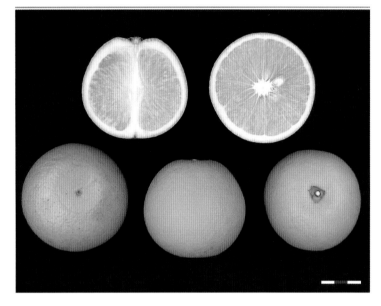

（3）明柳橙　果实长圆球形，橙黄色，大小为5.2～6.0cm×5.0～6.2cm。果顶有明显印圈，果面自蒂部至顶部有10余条柳纹，称为明柳。TSS 13.0％，酸0.5％～0.7％。种子10粒/果。味清甜，有香气。成熟期11月下旬至12月上旬。该品种适应性强，品质中上，渣较多。

(3) Mingliucheng　Fruit long-spheroid, yellow-orange, sizes in 5.2~6.0cm×5.0~6.2cm. Apex with conspicuous areole ring. Rind surface with about 10 willow leaf-like strips going vertically from base to apex (Mingliu means "pronounced willow leaf"). TSS 13.0%, acid content 0.5%~0.7%, 10 seeds per fruit. Fresh sweet, fragrant and less melting. Maturity period ranges from late November to early December. The variety is adaptable, of medium or higher quality.

21. 露德红夏橙

来源与分布：露德红夏橙又名红夏橙。美国佛罗里达州从伏令夏橙芽变中选育而成，我国于20世纪90年代初引进。

主要性状：性状与伏令夏橙相似。果实球形，果皮光滑，果肉色泽较普通伏令夏橙深，果汁色泽为深橙色。结果较早，果实较大，丰产性较好，成熟期与伏令夏橙相似。

21. Rohde Red Valencia Orange

Origin and Distribution: The variety is also named as Hongxia orange. It was selected from a bud sport of Valencia in Florida, USA. China introduced it in the 1990s.

Main Characters: Similar to Valencia orange. Fruit spheroid, rind smooth. Pulp color is darker than common Valencia orange. Juice deep orange colored. The variety is early fruiting with large fruit and relatively productive. Its maturing season is similar to Valencia orange.

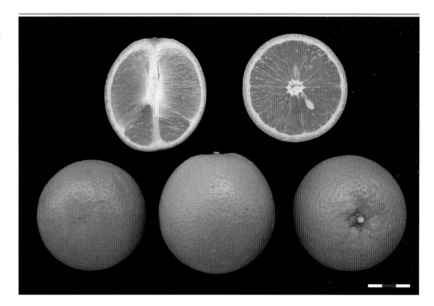

22. 蜜奈夏橙

来源与分布：南非从伏令夏橙芽变中选出。我国20世纪末从美国引进，现湖北、重庆有栽植。

主要性状：果实球形，橙色或深橙红，大小为6.8cm×7.2cm，果皮光滑，汁多风味优良。果皮薄，紧贴果肉，难于剥离。TSS 11.0%～12.0%，酸1.0%。果实少核或无核。肉质脆嫩、化渣、多汁，酸甜适度，风味浓郁。该品种丰产优质，成熟期比伏令夏橙早2周以上，也可挂树至伏令夏橙采收末期，果实既适宜鲜食又适宜加工。

22. Midknight Valencia Orange

Origin and Distribution: It was selected from a bud mutation of Valencia orange in South Africa. China introduced it from America at the end of last century and has some plantings in Hubei and Chongqing.

Main Characters: Fruit spheroid, orange to deep orange-red, sizes in 6.8cm×7.2cm.Rind smooth, thin and tightly adhered, difficult to peel.TSS 11.0%~12.0%, acid content 1.0%, seeds few or none. Pulp is crisp, melting and juicy with moderate sweet-sour taste and rich flavor. The variety is productive. Mature time is above 2 weeks earlier than Valencia, and is capable of holding on tree to the end of Valencia orange season. Fruit is of superior quality and is excellent for both fresh consumption and juicy processing.

23. 糖橙

来源与分布：原产埃及，我国1988年引入，现湖南、重庆有少量种植。

主要性状：树势旺，叶片长椭圆形。果实圆球形，橙红色，大小为6.1cm×6.6cm，果皮薄光滑，包着稍紧，难剥离。TSS 13.5%，酸0.55%，种子14粒/果。成熟期10月。该品种丰产，果肉脆嫩、多汁、化渣、纯甜。新近湖南从中选育出无核品系。

23. Tang Orange

Origin and Distribution: Originated in Egypt. It was introduced to China in 1988. It is currently cultivated a little in Hunan and Chongqing.

Main Characters: Tree vigorous. Leaf long-elliptical. Fruit spheroid, red-orange, sizes in 6.1cm×6.6cm; rind thin and smooth, tightly adherent, difficult to peel. TSS 13.5%, acid 0.55%, 14 seeds per fruit. Pulp crisp, tender, juicy, melting and pure sweet. Matures in October. The variety is productive. Recently some seedless lines have been selected in Hunan Province.

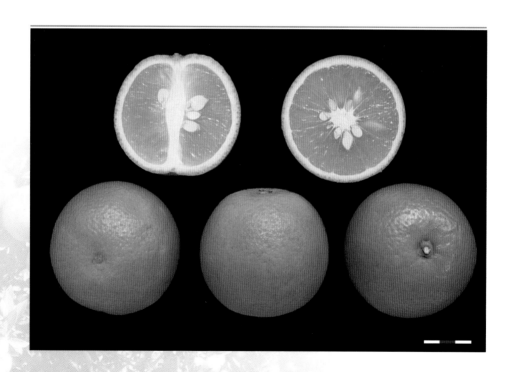

24. 桃叶橙

来源与分布：原产湖北省秭归县，20世纪50年代于甜橙实生树中选出。主产湖北省，广西等地有引种。

主要性状：树姿开张，树冠圆头形，枝条粗壮有短刺。春梢叶片大小为8.4～8.8cm×4.5～5.0cm，叶披针形，狭长似桃叶，故名桃叶橙。果实近圆形，橙红色，大小为5.3～5.8cm×5.1～5.4cm。果皮较难剥离。TSS 12.0%～16.0%，酸0.4%～0.8%。种子6～8粒/果。成熟期11月中下旬。该品种适应性强，果实品质优，较丰产。

24. Taoye Orange

Origin and Distribution: Taoye orange is derived from a chance seedling of common sweet orange in Zigui, Hubei Province in 1950s. It is mainly grown in Hubei, and introduced to Guangxi etc.

Main Characters: Tree spreading, spheroid, branches stout with short spines. Leaf sizes in 8.4~8.8cm×4.5~5.0cm, lanceolate like peach leaf (Taoye) and hence named Taoyecheng. Fruit nearly spheroid, red-orange, sizes in 5.3~5.8cm×5.1~5.4cm. somewhat difficult to peel. TSS 12.0%~16.0%, acid content 0.4%~0.8%, 6~8 seeds per fruit. Matures in middle to late November. The variety is quite adaptable, comparatively productive, with fruit of excellent quality.

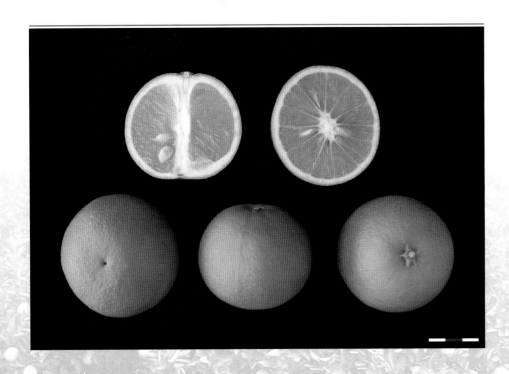

25. 先锋橙

来源与分布：先锋橙又名鹅蛋柑20号，原产重庆江津。重庆、四川、湖北有分布。

主要性状：树势和树性与锦橙基本相同，但枝条比锦橙硬，小刺稍多。果实短椭圆形，橙红色，大小为7.0cm×6.8cm，顶部稍宽，蒂部平或微凸，TSS 12.0%以上，酸1.0%。果肉酸甜，味浓，有香气。种子8粒/果，成熟期11月下旬至12月上旬。该品种丰产、稳产、果实品质优，鲜食加工果汁两用品种。砧木用枳和红橘。

25. Xianfengcheng

Origin and Distribution: Xianfengcheng is also named as Edangan No.20 and originated in Jiangjin of Chongqing. It is distributed in Chongqing, Sichuan and Hubei.

Main Characters: Tree vigor similar to Jincheng sweet orange but shoots stiffer with relatively more short spines. Fruit short-elliptical, orange red, 7.0cm×6.8cm, apex slightly broader, base truncate or slightly protruding, TSS above 12.0%, acid 1.0%, pulp sour-sweet, flavor rich and fragrant. 8 seeds per fruit. Maturity period ranges from late November to early December. The cultivar yields high and consistently. The high quality fruit is excellent for fresh consumption and juice processing. Common rootstocks are Trifoliate orange and Hongju.

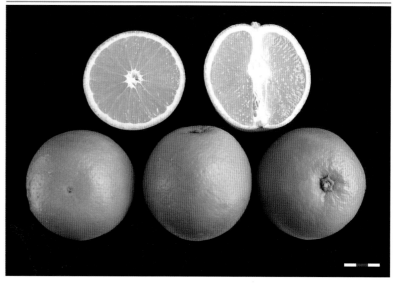

26. 香水橙

来源与分布：原产广东新会及广州近郊。广东、广西有栽培，湖南、浙江、福建等省有分布。

主要性状：树冠呈不规则圆头形，较开张，枝条茂密。春梢叶片大小为9.0～10.0cm×4.2～4.5cm，呈椭圆形，叶缘波状，锯齿较浅，翼叶稍大。花中大，完全花。果实圆球形或近椭圆形，橙黄色，大小为5.8～6.2cm×5.5～6.0cm，果顶柱点微凹，果蒂部微凹，有不明显沟纹，果皮难剥离。TSS 11.0%～12.0%，酸0.55%～0.65%。种子5～10粒/果。果肉细嫩，汁多。成熟期11月下旬至12月，可树上留果到翌年1月，风味更好。该品种适应性强，丰产性好，果实较耐贮藏，品质中等。

26. Xiangshui Orange

Origin and Distribution: Xiangshui orange originated in Xinhui, Guangdong Province and the suburb of Guangzhou. It is cultivated in Guangdong, Guangxi and distributed in Hunan, Zhejiang and Fujian etc.

Main Characters: Tree irregularly spheroid, relatively spreading, branches dense. Leaf 9.0~10.0 cm×4.2~4.5 cm in size, elliptical, margin sinuate with shallow crenate, petiole wing slightly large. Complete flower, medium sized. Fruit round to nearly ellipsoid, orange-yellow, sizes in 5.8~6.2cm×5.5~6.0cm, apex slightly concave, base slightly depressed with inconspicuous furrows, difficult to peel. TSS 11.0%~12.0%, acid content 0.55%~0.65%, 5~10 seeds per fruit. Pulp tender, juicy. Maturity period ranges from late November to early December. Fruit holds well on trees to next January and taste even better. The variety is adaptable and quite productive; fruit stores well, of medium quality.

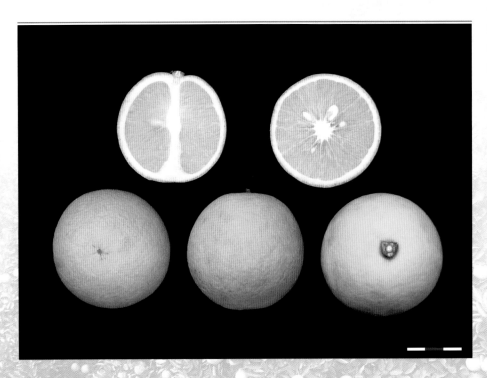

27. 新会甜橙

来源与分布：新会甜橙又名滑身仔、滑身橙，原产广东新会。广东、广西、福建有栽培，国内其他柑橘产区也有引种。

主要性状：长势中等，树冠半圆形，较开张，枝条较细。春梢叶片大小为8.2～9.4cm×3.3～3.8cm，呈菱形，叶缘波状，锯齿不明显，翼叶不明显。花较小，完全花。果实圆球形或长圆形，橙黄色至微橙红色，大小为6.0～6.4cm×5.4～5.7cm。果顶略平有印圈为本种特征，果皮不易剥离。TSS 13.0%～16.0%，酸0.1%～0.7%。种子6～8粒/果。果肉味甜汁多，细嫩，较化渣，有香味，品质上等。成熟期11月下旬至12月上旬。该品种适应性强，丰产稳产性好，是鲜食的优良品种。曾是广东的主栽培种。主要砧木为酸橘、三湖红橘。

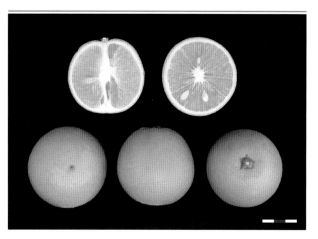

27. Xinhui Sweet Orange

Origin and Distribution: Xinhui sweet orange is also named as Huashenzai, Huashencheng, and originated in Xinhui, Guangdong Province. It is cultivated in Guangdong, Guangxi, Fujian, and introduced to other citrus producing areas in China.

Main Characters: Trees medium-vigorous, semi-spherical, relatively spreading, branch comparatively slender. Leaf 8.2~9.4cm×3.3~3.8cm in size, rhombus, margin sinuate, crenate not prominent, petiole wing inconspicuous. Complete flower, relatively small. Fruit spheroid to long-spheroid, orange-yellow to reddish orange, sizes in 6.0~6.4cm×5.4~5.7cm. Apex distinctively slightly flattened with an areole ring. Difficult to peel. TSS 13.0%~16.0%, acid content 0.1%~0.7%, 6~8 seeds per fruit. Pulp is sweet, juicy, tender, relatively melting and fragrant. Maturity period ranges from late November to early December. The variety is adaptable, consistently productive, and excellent for fresh consumption. It used to be a major cultivar in Guangdong Province. Sour tangerine and Sanhuhongju are commonly used as its rootstocks.

28. 溆浦长形甜橙

来源与分布：原产湖南溆浦，从普通甜橙芽变中选出。主产湖南。

主要性状：树势中等，树冠圆头形，枝梢刺少而短。春梢叶片大小为7.0～8.0cm×3.5～4.0cm，广披针形，叶缘波状。果实长圆形，橙色至深橙色，大小为5.5～6.0cm×6.0～7.0cm，果皮难剥离。TSS 11.0％，酸1.3％。成熟期11月下旬。该品种果大，无核或少核，肉细汁多，甜酸适中。

28. Xupuchangxing Sweet Orange

Origin and Distribution: Xupuchangxing sweet orange originated in Xupu, Hunan Province, as a bud sport of common sweet orange, and is now mainly grown in Hunan.

Main Characters: Tree medium-vigorous, spheroid, spines short and sparse. Leaf sizes in 7.0~8.0cm×3.5~4.0cm, broadly lanceolate, margin sinuate. Fruit long ellipsoid, orange to deep orange, sizes in 5.5~6.0cm×6.0~7.0cm, difficult to peel. TSS 11.0%, acid content 1.3%. Matures in late November. Fruit large in size, seedless or low-seeded. Pulp melting, juicy, moderate sweet sour.

29. 雪柑

来源与分布：雪柑又名雪橙、广橘。原产广东潮汕。广东、福建、台湾、广西、浙江等省（自治区）均有栽培。是我国橙类栽培历史比较悠久的一个品种。

主要性状：树势较强，树冠圆头形，枝条细长，较开张。春梢叶片大小为6.5～9.3cm×3.4～4.2cm，呈椭圆形。花中大，完全花。果形圆球形或广椭圆形，两端近对称，深橙黄色，大小为6.5～7.0cm×6.3～6.8cm，果顶圆，果皮难剥离。TSS 11.0%～13.0%，酸0.8%～1.1%。种子8～15粒/果。汁多有香气。成熟期11月下旬至12月上旬。该品种适应性较强，丰产稳产，风味浓，品质中上，耐贮运，为鲜食和果汁加工兼优品种。主要砧木为枳、三湖红橘等。

29. Xuegan

Origin and Distribution: Also named as Xuecheng or Guangju, Xuegan originated in Chaoshan, Guangdong Province. There are cultivations in Guangdong, Fujian, Taiwan, Guangxi and Zhejiang etc. It is one of varieties with a relatively longer history of cultivation among the Chinese sweet oranges.

Main Characters: Tree comparatively vigorous, spheroid, branch slender, long, relatively spreading. Leaf 6.5~9.3cm×3.4~4.2cm in size, elliptical. Complete flower, medium-large. Fruit spheroid to broad-elliptical, base and apex almost symmetrical, deep orange-yellow, sizes in 6.5~7.0cm×6.3~6.8cm, apex round, difficult to peel. TSS 11.0%~13.0%, acid 0.8%~1.1%. 8~15 seeds per fruit. Juicy and fragrant. Maturity period ranges from late November to early December. The variety is comparatively adaptable and productive, flavor rich, of medium or higher quality, tolerant to storage and shipment, and good for fresh consumption and juice processing. Common rootstocks are Trifoliate orange and Sanhuhongju.

（1）**大雪柑** 叶片较大。果实较大，广椭圆形，橙黄色，油胞大而疏，果实大小为7.0～7.5cm×6.8～7.3cm，果顶圆，果蒂微凹。TSS 11.0%～12.0%，酸0.8%～0.9%。种子10～15粒/果。汁多，有香味。

(1) Daxuegan Leaf and fruit are relatively larger. Fruit broad-elliptical, orange yellow, oil glands large and sparse, fruit sizes in 7.0~7.5cm×6.8~7.3cm, apex round, base slightly depressed. TSS 11.0%~12.0%, acid 0.8%~0.9%. 10~15 seeds per fruit, juicy and fragrant.

（2）零号雪柑　叶片比大雪柑略小，果实长圆形或球形，橙黄色，鲜艳有光泽，果实大小为6.5～7.2cm×6.2～7.0cm。TSS 11.0%～12.5%，酸0.9%～1.2%。种子8～12粒/果。肉质细嫩化渣，有香气，风味佳。

(2) Linghaoxuegan　Leaf slightly smaller than Daxuegan. Fruit long-spheroid or spheroid, orange-yellow, brilliant and glossy, fruit sizes in 6.5~7.2cm×6.2~7.0cm. TSS 11.0%~12.5%, acid 0.9%~1.2%. 8~12 seeds per fruit. Pulp tender, melting and fragrant with excellent flavor.

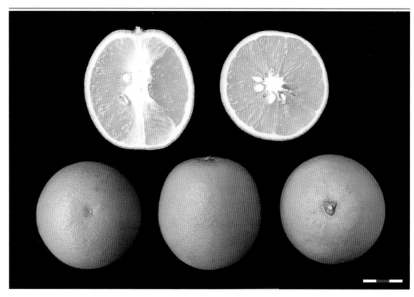

（3）小雪柑　叶片较小。果实较小，圆球形，橙黄色，油胞细密平生，果实大小为6.2～6.6cm×6.0～6.3cm，果顶圆。TSS 12.0%～13.0%，酸0.7%～0.8%。种子8～10粒/果。汁多，香味浓。

(3) Xiaoxuegan Leaf and fruit are relatively small. Fruit spheroid, orange-yellow, oil glands inconspicuous, small and dense. Fruit sizes in 6.2~6.6cm× 6.0~6.3cm, apex rounded. TSS 12.0%~13.0%, acid content 0.7%~0.8%. 8~10 seeds per fruit. Flesh is juicy with rich aroma.

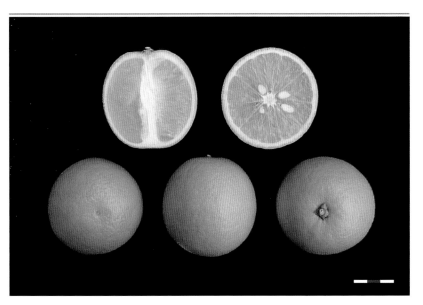

30. 中柑蜜橙

来源与分布：中柑蜜橙为长叶橙的早熟芽变选系，于2011年在中国农业科学院柑橘研究所育种圃中发现。2020年获得植物新品种权。

主要性状：树势较强，树冠自然圆头形。春梢叶片狭长，披针形。果实圆球形，单果重170g左右，果形指数0.93～0.98。果面较光滑，油胞稀疏，果皮厚0.38～0.45cm。果肉橙黄色，细嫩化渣，风味浓郁，有独特香气，果实少核。TSS11.8%～13.4%，固酸比13～17，每100ml果汁维生素C含量43.8～54.8mg，可食率65.1%～73.5%，鲜食加工兼宜。在重庆地区果实11月下旬成熟，果实耐贮藏，可留树保鲜至翌年3月上中旬。适宜在甜橙相似生态产区推广发展，尤其适宜在三峡库区作为鲜食与加工兼用品种推广。

30. Zhonggan Micheng

Origin and Distribution: Zhonggan Micheng is an early-season bud mutation of Changye sweet orange, found in Citrus Research Institute, Chinese Academy of Agricultural Sciences in 2011, and has got PVP of the Ministry of Agriculture and Rural Affairs in 2020.

Main Characters: Trees quite vigorous, naturally spheroid. Leaf long and narrow, lanceolate. Fruit spherical, fruit weight 170g, shape index 0.93~0.98. Rind smooth, with sparse oil glands, 0.38 to 0.45 cm thick. Flesh orange-yellow, tender and melting, flavor rich, aroma distinctive, few seeds. TSS 11.8%~13.4%, RTT (TSS/TA) 13~17, Vitamin C 43.8~54.8mg/100ml, edible portion 65.1%~73.5%, juice yield 51.0%~60.6%. Suitable for both fresh consumption and processing. In Chongqing, fruit matures in late November, stores well. Fruits can be kept fresh on trees until early-to mid-March of the following year. It is suitable for sweet oranges cultivation area, especially the Three Gorges Reservoir area.

（二）脐橙　Navel Oranges

1. 大三岛脐橙

来源与分布：原产日本，是华盛顿脐橙早熟芽变。我国1978年从日本引入，四川、重庆、广西、浙江、福建、广东等地有少量种植。

主要性状：果实圆球形或短椭圆形，橙黄到橙红色，大小为6.8～7.2cm×6.4～6.8cm，果顶圆，多为闭脐，TSS 10.0%～12.0%，酸0.5%～0.8%。果肉脆嫩、化渣、味清甜，品质上等，无核。成熟期11月中下旬。该品种树形矮小，丰产，皮薄较易裂果。

1. Omishima Navel Orange

Origin and Distribution: Omishima navel orange, originated in Japan as an early-maturity bud sport of Washington navel orange, was introduced into China in 1978. There are some plantings in Guangdong, Guangxi, Sichuan, Chongqing, Zhejiang and Fujian etc.

Main Characters: Fruit spheroid or short-ellipsoid, yellow-orange to red-orange, 6.8~7.2cm×6.4~6.8cm in size, apex rounded and mostly with closed navel, seedless. TSS 10.0%~12.0%, acid content 0.5%~0.8%. Flesh crisp, melting, fresh sweet and of superior quality. Matures in middle to late November. Tree dwarfed but productive. Rind thin but susceptible to splitting.

2. 奉节72-1脐橙

来源与分布：1972年重庆奉节县园艺场从四川江津园艺试验站一株甜橙砧的华盛顿脐橙后代中选出。主产重庆，四川、湖北等省有少量栽培。

主要性状：果实短椭圆形或圆球形，深橙至橙红色，大小为8.0cm×7.7cm，脐较小。TSS 11.0%～14.0%，酸0.7%～1.0%。肉细嫩，味清甜有香气，无核。成熟期11月下旬至12月上旬。该品种丰产、稳产，品质上等。

2. Fengjie 72-1 Navel Orange

Origin and Distribution: Fengjie 72-1 navel orange was selected from the offspring of a Washington navel orange tree on sweet orange rootstock obtained from the previous Jiangjin Horticultural Experiment Station, Sichuan, by Fengjie Horticultural Farm, Chongqing, in 1972. It is now mainly grown in Chongqing, and also cultivated to some extent in Sichuan and Hubei.

Main Characters: Fruit short-ellipsoid or spheroid, deep orange to reddish orange, sizes in 8.0cm×7.7cm, with a smaller navel. TSS 11.0%~14.0%, acid 0.7%~1.0%, seedless. Pulp is tender, fresh sweet and fragrant. Maturity period ranges from later November to early December. Yield high and consistent, quality superior.

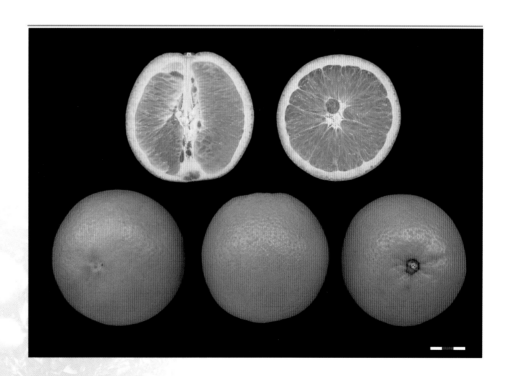

3. 奉晚脐橙

来源与分布：由奉节县脐橙研究所、华中农业大学等单位联合从奉园脐橙（奉节72-1）的芽变中选出。主产重庆奉节。

主要性状：果实卵圆形，深橙至深红色，大小为8.0cm×7.2cm，TSS 12.0%，酸0.7%。该品种着色较晚，成熟期比奉节72-1推迟2个月左右，1~2月成熟。丰产稳产，品质好。

3. Fengwan Navel Orange

Origin and Distribution: Fengwan navel orange is selected from a bud mutation of Fengjie 72-1 navel orange jointly by Fengjie Navel Orange Research Institute and Huazhong Agricultural University. It is mainly grown in Chongqing.

Main Characters: Fruit ovoid, deep orange to reddish-orange, sizes in 8.0cm×7.2cm, TSS 12.0%, acid 0.7%. Fruit breaks color is a little later and matures in January to February, two months later than Fengjie72-1 navel orange. Yields are high and consistent. Quality is fairly good.

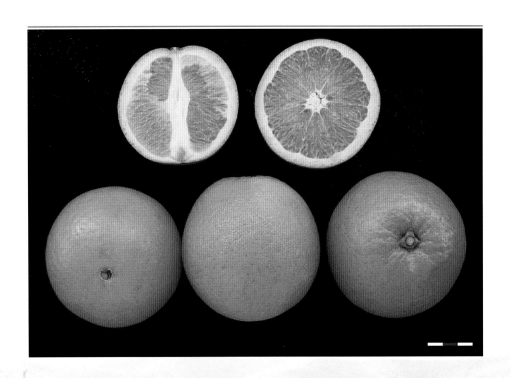

4. 福本脐橙

来源与分布：原产日本，华盛顿脐橙枝变。我国1981年从日本引入，重庆、四川等地有少量栽培。

主要性状：果实短圆形或球形，橙红色，多闭脐，蒂部周围有明显的放射沟。果皮中厚，较易剥离。TSS 12.0%，酸0.9%。肉质脆嫩，化渣，多汁，香气浓郁，品质优，无核。成熟期在南亚热带地区10月下旬可上市，最适上市时间是11月上旬。该品种产量中等，成熟早，果面光滑，色深而艳丽。

4. Fukumoto Navel Orange

Origin and Distribution: Fukumoto navel orange, originated in Japan as a limb sport of Washington navel orange, was introduced to China in 1981. There is some commercial production in Sichuan, Chongqing etc.

Main Characters: Fruit obloid or spheroid, red-orange, seedless. Base striped with prominent radial furrows, most with closed navel. Rind medium thick, easily peeling. TSS 12.0%, acid content 0.9%. Flesh crisp, melting and juicy. Fruit quality is excellent with rich aroma. It can be marketable in late October in southern subtropical regions. However, the best marketing time is early November. Fukumoto trees are medium-productive and early maturity. Rindis smooth, deep red with brilliant surface.

5. 福罗斯特脐橙

来源与分布：美国加利福尼亚州的品种，华盛顿脐橙的珠心系。我国1978年从美国引进。四川、重庆、湖北、江西等地有少量栽培。

主要性状：果实圆球形或椭圆形，橙红色，大小为7.1～7.4cm×7.1～7.6cm。TSS 11.0%～13.0%，酸0.8%～1.2%。果肉脆嫩、化渣、清甜，品质好。成熟期12月中下旬。该品种树势强壮，丰产稳产，可挂树贮藏至翌年1月底至2月初。

5. Frost Navel Orange

Origin and Distribution: It is a California variety, selected from nucellar seedling of Washington navel orange. Frost was introduced to China from America in 1978 and is currently planted to some extent in Sichuan, Chongqing, Hubei, Jiangxi.

Main Characters: Fruit spheroid or ellipsoid, reddish orange, sizes in 7.1~7.4cm×7.1~7.6cm. TSS 11.0%~13.0%, acid 0.8%~1.2%. Flesh crisp, melting, fresh sweet, and of good quality. Matures in later December. Tree vigorous; yield high and consistent. Fruit holds well on tree until later January or early February.

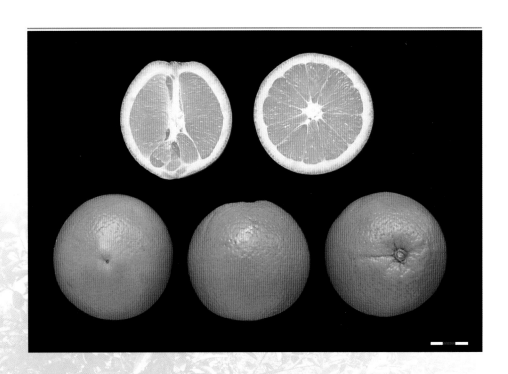

6. 赣南1号脐橙

来源与分布：赣南1号脐橙又名安远早脐橙，2007年在江西赣南安远县欣山镇教塘村纽荷尔脐橙园内发现的早熟芽变。2016年获得植物新品种权。在脐橙优势区均可栽培。

主要性状：树势中庸，树姿较开张，长势与纽荷尔脐橙相近。成枝力强，无刺。叶片卵圆形，叶色浓绿。成花能力强，以春梢结果为主。果实近圆球形，中等大小，果形指数1.06，单果重230g。成熟时果皮橙红色，果皮厚0.35cm，果肉细嫩化渣，口感好，多闭脐。TSS12.9%，酸0.7%，每100ml果汁维生素C含量69.8mg。与枳嫁接亲和性好，丰产、稳产，在安远地区10月中下旬成熟。

6. Ganan No.1 Navel Orange

Origin and Distribution: A bud mutation of Newhall navel orange, locally named Anyuan early navel orange, found in an orchard in Jiaotang, Xinshan, Anyuan County, Jiangxi Province in 2007. This variety got PVP of the Ministry of Agriculture in 2016. It can be cultivated in the area where navel oranges are grown popularly.

Main Characters: Tree moderately vigorous, canopy somewhat open, similar to Newhall navel orange. Branching capacity strong, thorns absent. Leaf blade ovate, dark green. Flowering ability is strong, and fruit twigs are mainly spring shoots. Fruit medium size, nearly spherical with fruit shape index 1.06, mostly with inner navel, average weight 230g. Mature fruit peels are orange-red, 0.35cm thick on average. Flesh tender and melting, taste good; TSS 12.9%, TA 0.7% and Vitamin C 69.8mg/100ml. Graft compatible well with Trifoliate Orange. Prolific and yield stable. Maturity in the mid to late October in southern Jiangxi.

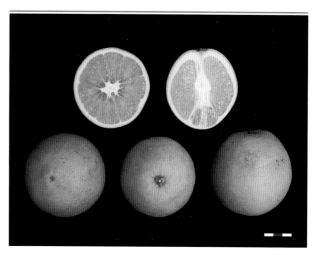

7. 赣南早脐橙

来源与分布：赣南早脐橙为纽荷尔脐橙芽变，2005年在江西省赣州市于都县禾丰镇一果园发现，2012年通过江西省农作物品种审定委员会审定。

主要性状：树势中庸，树冠为不规则圆头形，树姿开张。萌芽力弱、成枝力较强，枝梢节间长，无刺。叶色较浅，叶脉稀疏而突出，翼叶较大。成花力弱，坐果率高，结果母枝以早秋梢和春梢为主。果实中等偏大，单果重250g以上，近圆球形，多闭脐，果顶稍平；果皮油胞较大、稀、稍粗糙，果面橙红色。果肉细嫩化渣，TSS10.0%～13.0%，酸0.5%～0.6%，每100ml果汁维生素C含量 39.9～55.9mg，可食率82%左右。成熟期为10月中旬，赣南早脐橙与枳及其他脐橙嫁接均亲和，丰产、稳产。赣南早脐橙较耐溃疡病，但易感染枯枝型炭疽病，生产中应加强炭疽病的防治。适宜赣南—湘南—桂北脐橙优势产业带、福建三明市等同纬度地区及三峡库区低山（海拔450m以下）河谷地区，以冬季无霜冻的中亚热带产区最佳。

7. Gannan Zao Navel Orange

Origin and Distribution: A bud mutation of Newhall navel orange, found in an orchard in Hefeng Town, Yudu County, Jiangxi Province in 2005, and registered in 2012 in Jiangxi.

Main Characters: Tree medium vigorous, crown irregular rounded, spreading. Budding capacity weak, branching capability relatively strong, internode long, thorns absent. Leaf light color, veins sparse and prominent, petiole wing relatively large. Flowering ability weak, fruit-setting rate high. Fruits twigs are mainly early autumn shoots and spring shoots. Fruit medium-to-large size, average fruit weight above 250g, nearly spheroid, mostly with close navel and somewhat truncated apex. Oil glands large, sparse and rough. Rind orange-red. Pulp tender and melting, TSS 10.0%~13.0%, TA 0.5%~0.6%, Vitamin C 39.9~55.9 mg / 100ml, edible portion 82%. Maturity in mid-October in the south area of Jiangxi. Graft compatible with trifoliate orange and other navel oranges. Stably productive. Quite resistant to canker but susceptible to anthracnose, to which the prevention should be strengthened during the cultivation. It is suitable for the superior production belt of navel oranges in southern Jiangxi-southern Huan-northern Guangxi; Sanming City, Fujiang Province and areas of the same latitude; Low mountainous region (below 450m) and valley area in the Three Gorges Reservoir area; best in the mid-subtropical production area where winters are frost-free.

8. 红肉脐橙

来源与分布：红肉脐橙又名卡拉卡拉脐橙，是委内瑞拉从华盛顿脐橙中发现的芽变。我国于1990年从美国引进，目前在湖北、四川、重庆、广东等地有少量栽培。

主要性状：果实圆球形，橙黄色，大小为6.3～6.8cm×6.2～6.7cm，果皮难剥离，果顶多为闭脐。TSS 11.0%～12.0%，酸0.5%～0.8%，最突出的特征是果肉含番茄红素较多，呈均匀的红色。果汁多，味甜酸适中，无核。枝条木质部有正常和红色两种类型，红色木质部的类型果实偏小。该品种适于积温稍高地区种植。

8. Cara Cara Navel Orange

Origin and Distribution: Cara Cara navel orange, a mutation occurred on a Washington navel orange tree, was discovered in Venezuela. It was introduced into China from USA in 1990, and is currently planted to some extent in Sichuan, Chongqing, Hubei, Guangdong.

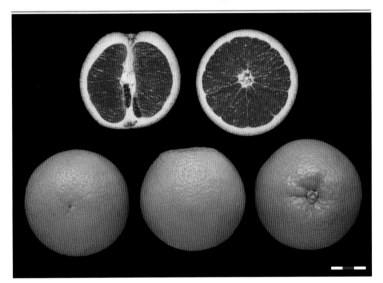

Main Characters: Fruit spheroid, yellow-orange, 6.3~6.8cm×6.2~6.7cm, difficult to peel, apex usually containing a closed navel. TSS 11.0%~12.0%, acid 0.5%~0.8%. Flesh is juicy, moderate sweet-sour and seedless with even pink color due to high content of lycopene. Xylem in some branches colored also in pink and fruit on such shoots is usually smaller. The variety adapts better in areas with higher accumulated temperature.

9. 华盛顿脐橙

来源与分布：华盛顿脐橙是14世纪葡萄牙人从中国引去甜橙，后转入巴西，在巴西Bahia从中选出无核有脐类型，称为Bahia橙；1870年引入美国，并定名为华盛顿脐橙。我国20世纪30年代从美国引进，现四川、重庆、湖北、湖南、江西、广东、广西等地均有栽培。

主要性状：树冠扁圆形或圆头形，开张，大枝粗长，披垂。春梢叶片大小为9.5～9.8cm×5.5～5.8cm，椭圆形，叶缘为全缘，翼叶小。花稍大，雄性不育。果实长圆球形或倒卵形，橙色或橙红色。单果重250g左右，果顶有脐，脐孔开或闭。果皮较难剥离。TSS 10.0%～14.0%，酸0.5%～1.0%。肉质脆嫩化渣，酸甜适中，无核。成熟期11月下旬至12月。容易发生芽条变异，世界上多数脐橙品种是从其芽变中选出。

9. Washington Navel Orange

Origin and Distribution: In 14th century the Portuguese first introduced common sweet orange from China, which was eventually brought to Brazil. A seedless navel orange mutation was selected from these orange trees in Bahia, a Brazilian city, and hence named Bahia orange. It was named Washington navel orange after its introduction to USA in 1870. China introduced it from USA in 1930s. Plantings of Washington navel orange can be found in Sichuan, Chongqing, Hubei, Hunan, Jiangxi, Guangdong, Guangxi and some other provinces.

Main Characters: Tree obloid or spheroid, spreading. Large branches thick and long, drooping. Leaf size 9.5~9.8cm×5.5~5.8cm, elliptic, margin entire, petiole wing small. Flower somewhat large, male sterile. Fruit ellipsoid to obovoid, rind orange to red-orange, weights about 250g. Apex with a navel, open or closed. Comparatively difficult to peel. TSS 10.0%~14.0%, acid 0.5%~1.0%. Pulp is crisp, melting and seedless with moderate sweet-sour flavor. Maturity period ranges from late November to December. Bud sports are prone to occur, giving rise to most navel orange varieties in the world.

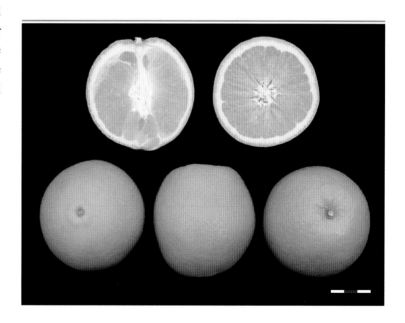

10. 崀丰脐橙

来源与分布：湖南农业大学与湖南省新宁县农业局从华盛顿脐橙选出，原代号7904，属华盛顿脐橙芽变系，2004年通过湖南省农作物品种审定委员会审定。主产湖南，广西、江西、湖北等地有少量栽培。

主要性状：果实圆形或椭圆形，橙红色，大小与华盛顿脐橙类似，果脐小，多闭脐。果蒂部周围有短浅放射沟。TSS 11.0%～13.0%，酸0.8%～1.1%。无核。成熟期11月中下旬至12月上旬。该品种适应性强，丰产性好，果实品质好，耐贮藏。

10. Langfeng Navel Orange

Origin and Distribution: Langfeng navel orange, previously numbered as 7904, was selected from a bud sport of Washington navel orange by Hunan Agriculture University and the Agricultural Bureau of Xinning, Hunan, and was registered as new variety by Hunan Crop Cultivar Registration Committee in 2004. It is mostly grown in Hunan and limitedly in Guangxi, Jiangxi, and Hubei etc.

Main Characters: Fruit spheroid to ellipsoid, red-orange, similar to Washington navel orange in size, navel small and mostly closed. Base with short shallow radial furrows. TSS 11.0%~13.0%, acid content 0.8%~1.1%, seedless. Maturity period ranges from later November to early December. The tree is very productive and adaptable. The fruit is of fine quality with good storage life.

11. 伦晚脐橙

来源与分布：伦晚脐橙为华盛顿脐橙芽变，20世纪50年代澳大利亚从华盛顿脐橙中选育的芽变品种。21世纪初由华中农业大学分别从美国和澳大利亚引进。2010年通过湖北省农作物品种审定委员会认定。

主要性状：树势强健，树形为不规则的自然圆头形，较开张。萌芽力中等，成枝力强，新梢具浅刺。成花能力强，结果母枝以春梢或早秋梢为主。果实长椭圆形，完全闭脐，无籽，果顶微凸出，果面光滑、橙黄微红，单果重260g。果肉细嫩化渣，TSS12.0%以上，酸0.7%，每100ml果汁维生素C含量38.9mg，可食率45.5%。在三峡库区翌年3月中下旬果实完全成熟。

伦晚脐橙与枳、温州蜜柑及其他脐橙均嫁接亲和，丰产、稳产，适宜三峡库区海拔350m以下河谷地区或年均温达到18℃以上、冬季冷凉但无霜冻的地区。

11. Lane Late Navel Orange

Origin and Distribution: Lane late navel orange, a bud mutation of Washington navel orange selected in Australia in the 1950s, was introduced by Huazhong Agricultural University at the beginning of the 21st-century from the USA and Australia respectively; this variety got registration in Hubei Province in 2010.

Main Characters: Strong tree vigor, irregular natural spherical crown, opening canopy, medium budding capacity, strong shoot formation ability, new twigs with tiny thorns. Flowering ability is strong, and the fruiting twigs are mainly spring or early autumn shoots. Fruit oblong, closed navel, seedless, apex slightly convex. Rind smooth, orange-yellow to light red in color; fruit weight 260 g, pulp tender and melting, TSS 12.0%, TA 0.7% Vitamin C 38.9mg/100ml, and edible portion 45.5%. Fruits mature in late March next year in the Three Gorges Reservoir area. Graft compatible to trifoliate orange, Satsuma mandarin and other navel oranges, prolific and yield stable. It can be grown in the Three Gorges Reservoir area where the altitude is lower than 350m or area with an average annual temperature above 18 ℃, cool but frost-free in the winter.

12. 罗伯逊脐橙

来源与分布：罗伯逊脐橙又名鲁宾逊脐橙、罗脐，原产美国，从华盛顿脐橙芽变中选育而成。我国1938年从美国引入，四川、重庆、湖南、湖北、浙江等地有栽培。

主要性状：果实近球形或倒卵形，橙红色，大小为6.5～6.7cm×6.2～6.5cm，果顶浑圆或微凹，开脐或闭脐，稍光滑。TSS 11.0%～13.0%，酸0.6%～1.0%。果肉脆嫩化渣，味较浓，微香，品质优，无核。成熟期11月上中旬。该品种树势中等，矮化开张。较丰产稳产，坐果率较华盛顿脐橙高，幼树脐黄稍多。树干有瘤状突起。

12. Robertson Navel Orange

Origin and Distribution: Robertson navel orange is a bud sport of Washington navel orange selected in USA. It was introduced to China from America in 1938, and limitedly cultivated in Sichuan, Chongqing, Hubei, Hunan, Zhejiang.

Main Characters: Fruit subspheroid to obovoid, reddish orange, sizes in 6.5~6.7cm×6.2~6.5cm. Apex rounded or slightly depressed with open or closed navel. Rind slightly smooth. TSS 11.0%~13.0%, acid 0.6%~1.0%. Flesh crisp, melting, rich flavor with mild aroma, and of good quality. Matures before late November. Tree medium-vigorous, dwarfed and spreading. Trunks are covered with lumps. Yields are good and consistent. Ratio of fruit-setting is higher than that of Washington navel orange. Navel yellowing problem is more pronounced on young trees.

13. 梦脐橙

来源与分布：美国佛罗里达州从华盛顿脐橙芽变中选出。我国于20世纪90年代引进，湖北有少量栽培。

主要性状：果实圆球形，橙黄色，光滑，多闭脐，果皮稍难剥离。TSS 11.0%～12.0%，酸0.6%～0.9%。无核，肉质柔软，风味浓。成熟期10月下旬至11月初。该品种丰产性好。

13. Dream Navel Orange

Origin and Distribution: Dream navel orange, selected from a bud sport of Washington navel orange in Florida, USA, was introduced to China in 1990s and has limited plantings in Hubei Province.

Main Characters: Fruit spheroid, yellowish orange, smooth, usually with closed navel, slightly difficult to peel. Seedless. TSS 11.0%~12.0%, acid content 0.6%~0.9%. Pulp soft and rich in flavor. Matures in late October to early November. Very productive.

14. 奈维林娜脐橙

来源与分布：西班牙主栽品种之一，华盛顿脐橙芽变。我国1979年从西班牙引进，四川、重庆、湖北、福建、湖南、江西、浙江、广西及广东有少量栽培。

主要性状：果实长椭圆形或倒卵形，深橙或橙红色，大小为7.9～8.3cm×8.8～9.3cm，果顶圆钝，基部较窄，常有短小沟纹。TSS 11.0%～13.0%，酸0.6%～0.8%。果肉脆嫩，化渣，品质优。无核。成熟期11月上中旬。该品种树势偏弱，产量不稳定，果实可适当在树上挂果贮藏。

14. Navelina Navel Orange

Origin and Distribution: Navelina navel orange, originated from a bud sport of Washington navel orange, is one of the most popular varieties in Spain. It was introduced into China from Spain in 1979. Currently, it is planted to some extent in Sichuan, Chongqing, Hubei, Fujian, Hunan, Jiangxi, Zhejiang, Guangxi and Guangdong.

Main Characters: Fruit oblong or obovate, deep orange or red-orange, 7.9~8.3cm×8.8~9.3cm, apex obtuse or rounded; base slightly narrow and usually with faint furrows. TSS 11.0%~13.0%, acid 0.6%~0.8%. Flesh crisp and melting. Fruit quality is excellent without seeds. Matures in early to middle November. Trees are not very vigorous. Yields are not consistent. Fruit could hold on tree for some time.

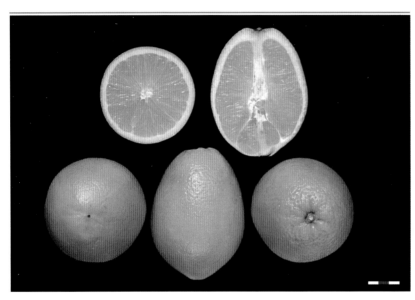

15. 纽荷尔脐橙

来源与分布：纽荷尔脐橙是华盛顿脐橙早熟芽变种，我国1980年从美国、西班牙同时引进。主产江西、湖北、湖南、四川、重庆、广东、广西、福建，其他柑橘主产区也有少量栽培，是目前中国脐橙中栽培面积最大的品种。

主要性状：树势较强，树姿开张，树冠圆头形，枝条粗长，披垂，有短刺。春梢叶片大小为9.4～9.8cm×5.5～5.9cm，椭圆形，叶色深绿色，叶缘为全缘，翼叶小。花稍大，花粉败育。果实椭圆形，顶部稍凸，脐多为闭脐，蒂部有5～6条放射沟纹，橙红至深红橙色，大小为6.5～7.4cm×6.7～7.8cm，果皮难剥离。TSS 12.0%～13.5%，酸0.9%～1.1%。无核。果肉汁多化渣有香味，品质上等。该品种丰产性好，对硼比较敏感，容易出现缺硼症状。

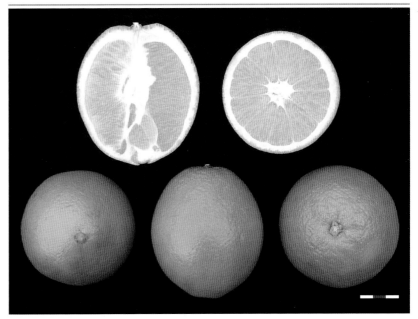

15. Newhall Navel Orange

Origin and Distribution: Newhall navel orange was selected from an early-maturing bud sport of Washington navel orange. China introduced it from both USA and Spain in 1980. As the most planted variety, Newhall has gained the highest popularity among the existing navel orange varieties in China. Its production is centered in Jiangxi, Hubei, Hunan, Sichuan, Chongqing, Guangdong, Guangxi and Fujian. There are introductions in some other citrus production areas.

Main Characters: Tree vigorous, spreading, spheroid, branches thick and long, drooping. Leaf sizes 9.4~9.8cm×5.5~5.9cm, elliptic, dark green, margin entire, with small petiole wing. Flower slightly large, pollen sterile. Fruit oblate; apex slightly protruding, with closed navel; base commonly with 5~6 radial furrows; red-orange to deep orange red, fruit sizes 6.5~7.4cm×6.7~7.8cm; difficult to peel. TSS 12.0%~13.5%, acid 0.9%~1.1%, seedless. Pulp is juicy, melting, fragrant, and of superior quality. The variety is very productive but sensitive to boron availability, easy to develop boron deficiency symptom.

16. 朋娜脐橙

来源与分布：美国加利福尼亚州品种，华盛顿脐橙芽变。我国1980年引入，四川、重庆、湖北、湖南、广西等地有栽培。

主要性状：果实圆球形或卵圆形，橙黄色或橙红色，大小为7.9cm×8.1cm，TSS 11.0%～14.0%，酸0.9%，无核。成熟期11月中旬前后。该品种早结丰产性好，优质早熟，但裂果落果较严重，开脐果比例较大。树干有瘤状突起。

16. Skaggs Bonanza Navel Orange

Origin and Distribution: It is a Californian variety, selected from a bud sport of Washington navel orange, introduced to China in 1980, and cultivated in Sichuan, Chongqing, Hubei, Hunan, Guangxi etc.

Main Characters: Fruit spheroid or ovoid, yellowish orange to reddish orange, sizes in 7.9cm×8.1cm, TSS 11.0%~14.0%, acid 0.9%, seedless. Fruit quality is good though most with open navel and prone to splitting. This cultivar is quite productive and maturing around middle November. The trunk has distinctive lumps.

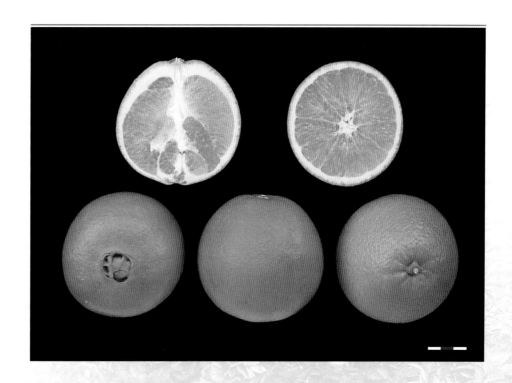

17. 黔阳冰糖脐橙

来源与分布：黔阳冰糖脐橙是从普通冰糖橙芽变选育的新品种，2000年在湖南省洪江市沙湾乡果园发现。2015年获得植物新品种权。

主要性状：树势中庸，树冠圆头形，树姿开张，成枝力强，刺少。成花力强，结果母枝以春梢或早秋梢为主。果实圆球形，单果重190～220g，果脐闭合，无种子。果面橙色，较光滑，囊壁较薄，果肉质地脆嫩，化渣性好。香气清淡。TSS12.7%，酸0.7%，每100ml果汁维生素C含量61.6mg，出汁率52%，可食率65%。黔阳冰糖脐橙与枳、温州蜜柑及其他甜橙均嫁接亲和，丰产、稳产，在怀化地区11月下旬完全成熟。适宜年均温达到17.5℃、年积温达到5 500℃以及最低温高于–2℃的甜橙优势栽培区。

17. Qianyang Bingtang Navel Orange

Origin and Distribution: A bud mutation of common Bingtangcheng, was found in 2000 in an orchard in Shawan, Hongjiang City, Hunan Province. This variety got PVP of the Ministry of Agriculture in 2015.

Main Characters: Tree medium vigorous, crown round and open, with strong branching capacity and few thorns. Flowering ability is strong, and fruiting twigs are mainly spring shoots or early autumn shoots. Fruit spherical, single fruit weight 192~220g, with inner navel, seedless; Rind orange, surface smooth, segment membrane thin; Flash crisp, tender and melting; Aroma light. TSS 12.7%, TA 0.7%, Vitamin C 61.6mg/100ml, juicing rate 52% and edible portion 65%. Qianyang Bingtang Navel Orange is graft-compatible with Trifoliate Orange, Satsuma mandarin and other oranges. Prolific and production stable. Full maturity in late November in Huaihua, Hunan. It is Suitable for sweet orange production area where an average annual temperature is above 17.5 ℃, an annual accumulated temperature of 5 500℃, and a minimum temperature above –2℃.

18. 青秋脐橙

来源与分布：青秋脐橙为眉红脐橙的早熟芽变，2006年在中国农业科学院柑橘研究所育种圃眉红脐橙园中发现。2018年获得植物新品种权。

主要性状：树势中庸，树冠自然圆头形。枝条粗壮，节间较短，坐果率高，果实卵圆形或长椭圆形，单果重270g，果形指数1.18。果顶较圆，98%以上为闭脐。果皮橙红色，较光滑，油胞较小、微凸，皮厚0.55cm，易剥离。果肉橙黄色，脆嫩、化渣、无核。TSS11.4%～14.2%，酸0.4%～0.9%，固酸比18.6∶1，每100ml果汁维生素C含量53.8～61.8mg，可食率68.9%～73.5%，出汁率48.8%～55.5%。丰产稳产，在重庆成熟期10月上中旬，比纽荷尔脐橙早45～50d，果实耐贮藏。适宜在脐橙相似生态产区推广发展，尤其适宜在三峡库区作为鲜食品种推广。

18. Qingqiu Navel Orange

Origin and Distribution: Qingqiu navel orange is an early-season mutation of Meihong navel orange, found in the navel orange orchard of Citrus Research Institute, Chinese Academy of Agricultural Sciences in 2006. It is registered in the Ministry of Agriculture and Rural Affairs in 2018.

Main Characters: Tree medium in vigor, crown naturally spheroid. Twigs thick and strong, internode short, fruit setting rate high, prolific and yield stable. Fruit ovoid or oblong, average weight 270g, fruit shape index 1.18. Apex relatively round, 98% with closed navel. Rind orange-red, smooth, 0.55 cm thick, oil glands small and slightly convex, easy to peel. Pulp orange-yellow, crisp and tender, melting, seedless. TSS 11.4%~14.2%, TA 0.4%~0.9%, Vitamin C 53.8~61.8mg/100ml, edible portion 68.9%~73.5%, juicing rate 48.8%~55.5%. Fruit ripening in early to mid-October in Chongqing, 45~50 days earlier than that of Newhall Navel Oranges. It is suitable for navel orange production area, especially as fresh consumption cultivar in the Three Gorges Reservoir area.

19. 清家脐橙

来源与分布：原产日本，是华盛顿脐橙早熟芽变。我国1978年从日本引入，四川、重庆有少量栽培，湖北、湖南、江西、福建、广东、广西、云南、贵州有引种。

主要性状：果实圆球形或近椭圆形，深橙色，大小为7.0～7.4cm×6.8～7.8cm，多为闭脐。TSS 11.0%～12.5%，酸0.7%～0.9%。果肉柔软，多汁，化渣，品质优，无核。成熟期11月上中旬。该品种适应性强，早结丰产，幼果生长快易裂果。

19. Seike Navel Orange

Origin and Distribution: Seike navel orange, originated in Japan as a early-maturity bud sport of Washington navel orange, was introduced to China in 1978. There are some commercial productions in Sichuan and Chongqing and introductions in Hubei, Hunan, Jiangxi, Fujian, Guangdong, Guangxi, Yunnan and Guizhou.

Main Characters: Fruit spheroid or nearly ellipsoid, deep orange color, 7.0~7.4cm×6.8~7.8cm in size and seedless. Most fruits have closed navel. TSS 11.0%~12.5%, acid content 0.7%~0.9%. Pulp is tender, juicy and melting. Fruit quality is excellent. Matures in early to middle November. The variety is quite adaptable, early fruiting and very productive. Young fruit grows fast, and is prone to split.

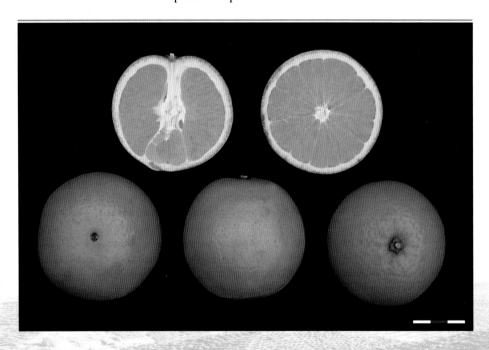

20. 园丰脐橙

来源与分布：园丰脐橙为华盛顿脐橙的优良变异，1993年在湖南省郴州市资兴县坪石乡华盛顿脐橙园发现。2011年通过品种登记。适合在现有纽荷尔脐橙种植区推广。

主要性状：树势强，树冠较开张，萌芽力中等，成枝力强。结果母枝以春梢、早秋梢为主。果实近圆球形，多闭脐，果面光滑、橙红色，单果重264g，大小为6.9～8.7cm×7.5～8.4cm，果形指数0.99～1.06，果皮厚0.48～0.61cm，无核，可食率69.6%～73.4%，TSS 12.0%以上，总糖9.7%～11.1%，酸0.7%～1.1%，每100ml果汁维生素C含量49.9～68.4mg，固酸比13.4～18.0。风味甜酸适度、化渣，品质优良，耐贮性强。坐果率高，丰产稳产，在湖南成熟期12月上中旬，比纽荷尔脐橙晚15～20d。

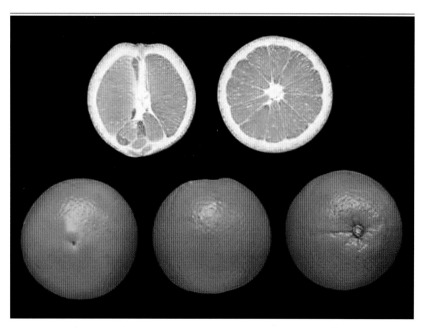

20. Yuanfeng Navel Orange

Origin and Distribution: Yuanfeng navel orange is an excellent variation found in Washington Navel Orange orchard in Pingshi Township, Zixing County, Chenzhou of Hunan in 1993. The variety was registered in 2011. It is suitable for the area where Newhall Navel Orange is grown.

Main Characters: Tree vigorous, crown relatively spreading, medium budding rate, and strong branching ability. Mainly spring and early autumn shoots bear fruits. Fruit is nearly spheroid, with most closed navel and smooth orange-red peel. Average fruit weight 264 g, width 6.9~8.7cm, height 7.5~8.4cm, fruit shape index 0.99~1.06; peel thickness 0.48~0.61cm, seedless, edible portion 69.6%~73.4%, TSS 12.0%, total sugars 9.7%~11.1%, titratable acidity (TA) 0.7%~1.1%, Vitamin C 49.9~68.4mg/100ml in juice. TSS/TA ratio (RTT) 13.4~18.0, moderately sweet, sour and melting pulp. With excellent quality and good storability. High fruit setting rate and stably prolific. Fruits mature in early to mid-December 15~20 days later than Newhall navel orange in Hunan.

21. 早红脐橙

来源与分布：早红脐橙由华中农业大学和秭归县合作选育，源自湖北秭归县一个温州蜜柑高接罗伯逊脐橙的果园。在湖北秭归，早红脐橙也称九月红，可留树保鲜至翌年3月，果肉十分细嫩，民间也称果冻橙。经过分析证明，该变异为温州蜜柑和脐橙嫁接形成的周缘嵌合体，L1层为温州蜜柑，L2层为脐橙，性状十分稳定。2008年获得植物新品种权。

主要性状：树势类似温州蜜柑，中等偏弱，成花力较强。叶片大小为3.2cm×7.9cm，叶形指数2.5，介于温州蜜柑和脐橙之间，偏向温州蜜柑。开花结果习性与脐橙相似，以有叶花序枝为主。花药无花粉。花期与温州蜜柑相同，果实近圆球形，偶尔有脐痕，单果重170g左右；果皮似橙，具有脐橙果皮的特色和香味；果肉似温州蜜柑，为橙红色，含有β-隐黄质；化渣性好，无籽。TSS11.8%～13.7%，酸0.49%～0.53%，每100ml果汁维生素C含量43mg；可食率为35.6%～44.7%。抗性与温州蜜柑类似。成熟期10月上旬。没有冻害的地方可以留树保鲜到翌年3月。在10月温度较高时采收，容易出现油斑病。该品种适宜温州蜜柑产区种植，无霜冻地区可以适度晚采。

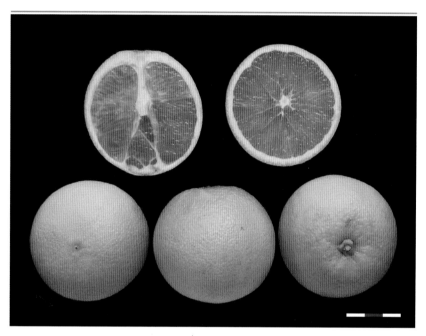

21. Zaohong Navel Orange

Origin and Distribution: Zaohong navel orange is an extra early navel orange variety, jointly selected by Huazhong Agricultural University and Zigui County, Hubei Province from an orchard where Robertson navel orange was grafted on Satsuma mandarin. Locally known as Jiuyuehong (September Red), fruits can be kept on tree after ripening until March in the following year, also known as Jelly orange for its very soft and tender jelly-like pulp. Genetic analysis verified it is a periclinal chimera generated from the grafted union of navel orange and satsuma mandarin. Its L1 is from satsuma mandarin, and the L2 from navel orange. It got PVP of the Ministry of Agriculture in 2008.

Main Characters: Tree vigor moderate-to-weak, similar to Satsuma mandarin, flowering ability strong. Leaf size between Satsuma mandarin and navel orange, but bias to mandarin, with an average size of 3.2cm×7.9cm, leaf shape index 2.5. Flowering and fruiting habits similar to navel oranges, mainly with leafy inflorescence like sweet orange. Pollen abortive. Flowering at the same time as Satsuma. Fruit nearly spherical, occasionally with the trace of navel, weight around 170 g; Rind with aroma and other properties similar to sweet orange; Pulp similar to Satsuma, orange-red colored, containing β-cryptoxanthin; good mastication, seedless, TSS 11.8%~13.7%, TA 0.49%~0.53% and Vitamin C 43 mg / 100ml, edible portion 35.6%~44.7%. Resistance similar to Satsuma. Maturity in early October. Fruits can hang on the trees until March of next year. When harvesting meets high temperature in October, the physiologic disorder of oil glands will happen, namely oleocellosis. It is suitable for satsuma producing areas. It can also be cultivated as a late-season variety in the frost-free region.

22. 宗橙脐橙

来源与分布： 宗橙系伦晚脐橙的果皮棕色芽变，由华中农业大学和秭归县共同选育。芽变发现于湖北秭归县归州镇。经过多年的研究表明，该变异性状稳定，特色明显，2020年获得植物新品种权。

主要性状： 植物学特性类似伦晚脐橙，树势强，树姿披垂，树冠半圆头形或圆头形，枝条较稀疏粗壮、具小刺。叶片阔披针形，大而肥厚，叶色浓绿。开花和成熟期与伦晚相似，在秭归4月底开花，成熟期为翌年3~4月。果实近圆球形，果实大小为8.3cm×7.6cm，单果重180~240g，脐小，多闭合，果皮为棕色，果皮厚度为0.6cm，无籽，果肉橙色，着色均匀，肉质脆嫩，富香气。TSS12.1%，酸0.5%，每100ml果汁维生素C含量51mg；比伦晚多了淡淡的清香味。该芽变形成棕色果皮的原因是滞绿基因（$CsSGR$）的两个碱基发生了突变，导致基因功能部分缺失，在果实成熟季节不能驱动叶绿素降解，在果皮中同时存在叶绿素和类胡萝卜素，形成了棕色的果皮外观。由于该基因为组成型表达，叶片落地后一段时间仍然保持绿色。该品种作为特殊的品种，适合无霜冻能种植晚熟脐橙的区域。

22. Zongcheng Navel Orange

Origin and Distribution: Zongcheng navel orange, a bud mutation with brown skin from Lane late navel orange, was jointly selected by Huazhong Agricultural University and Zigui County. It was found in Guizhou Town, Zigui County, Hubei Province. Long-term research shows that variation of peel color is genetically stable and distinctive. It has got PVP of the Ministry of Agriculture and Rural Affairs since 2020.

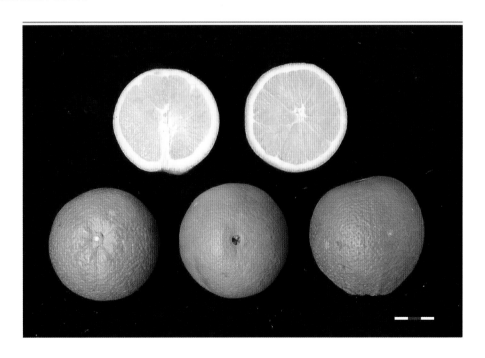

Main Characters: The plant characteristic of this variety is similar to Lane Late Navel Orange; tree vigorous, weeping, crown semi-spheroid or spheroid. Twigs sparse and stout, thorns small; Leaf blade lanceolate, large and thick, dark green. The flowering and mature season was similar to that of Lane Late, flowering at the end of April in Zigui, and the mature period was from March to April next year. Fruit nearly spherical, medium size in 8.3cm×7.6cm, weight 180~240g, navel small and mostly closed. Rind brown color, 0.6cm in thickness. Seedless, pulp color uniformly orange, crisp and tender, aroma fresh and richer than Lane Late. TSS 12.1%, TA 0.5% and Vitamin C 51 mg/100ml. Multi-omics study uncovered that the *CsSGR* gene is mutated, resulting in partial loss of gene function, as the chlorophyll degradation in the flavedo cells cannot be driven during maturity, but can still enhance the carotenoid biogenesis. The co-existence of green chlorophyll and orange carotenoids results in the brown peel. Fallen leaves stay green due to the constitutive expression of the mutated gene. As a special variety, this variety is suitable for areas where late-maturing navel oranges can be planted without frost in the winter.

（三）血橙　Blood Oranges

1. 红玉血橙

来源与分布：红玉血橙又名路比血橙，原产地中海地区。我国引入后四川有少量栽培，长江三峡地区、浙江等地有引种。

主要性状：树势中等，树冠圆头形，紧凑，半开张，枝梗有短刺。叶片长卵圆形，较小。果实扁圆形或球形，深红色，并带紫色斑纹，单果重150～200g，TSS 10.0%～11.0%，酸1.0%～1.1%。果汁多，甜酸适中，具玫瑰香味。种子10～13粒/果。成熟期1～2月，较丰产。

1. Ruby Blood Orange

Origin and Distribution:　Ruby blood orange originated in the Mediterranean area. There is limited production in Sichuan after its introduction into China. It was also introduced to the Three Gorges area of the Yangtze River and Zhejiang Province.

Main Characters: Tree medium-vigorous, compact spheroid, semi-spreading, branches with short spines. Leaf long-ovate, relatively small. Fruit oblate to globose, deep red with purple stripes; weight about 150~200g. TSS 10.0%~11.0%, acid content 1.0%~1.1%; flesh juicy with moderate sweet-sour flavor and rose-like aroma. 10~13 seeds per fruit. Matures in January to February. The variety is relatively productive.

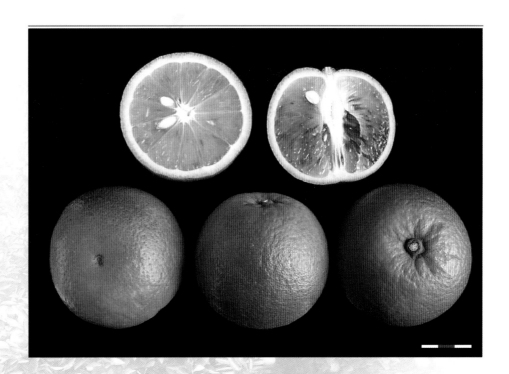

2. 脐血橙

来源与分布：原产西班牙，我国从摩洛哥引进，四川、浙江、广东、广西等地有少量栽植。

主要性状：树势中等，树冠圆头形，枝梢较密，具有短刺。春梢叶片大小为7.0～7.6cm×3.0～3.2cm，呈椭圆形，叶缘波状。花中大，完全花。果实椭圆形，橙黄色，光滑，单果重180～200g，果顶微有乳凸，花柱常宿存，果皮难剥离。TSS 11.0%～12.0%，酸0.9%～1.0%，无核。成熟期1～2月，果形美观，品质中上。成熟时果皮、果肉有红斑。该品种适应性强，丰产性好，果实耐贮运。可采用红橘作砧木，如用枳砧，裂皮病比较严重。

2. Washington Sanguine Blood Orange

Origin and Distribution: Washington sanguine blood orange, originated in Spain, was introduced to China from Morocco and is cultivated limitedly in Sichuan, Zhejiang, Guangdong and Guangxi etc.

Main Characters: Tree medium-vigorous, spheroid, branches comparatively dense with short spines. Leaf 7.0~7.6cm×3.0~3.2cm in size, elliptical, margin sinuate. Complete flower, medium-large. Fruit ellipsoid, orange-yellow, rind smooth, difficult to peel; weight 180~200g; apex with a small nipple and sometimes a persistent style. TSS 11.0%~12.0%, acid content 0.9%~1.0%, seedless. Matures in January to February. Fruit quality is medium or above with beautiful appearance. Pulp and pericarp speckled with red after reaching ripening. The variety is quite adaptive, productive. Fruit stores and ships well. Hongju can be used as its rootstock. Citrus exocortis may be a problem when using Trifoliate orange as its rootstock.

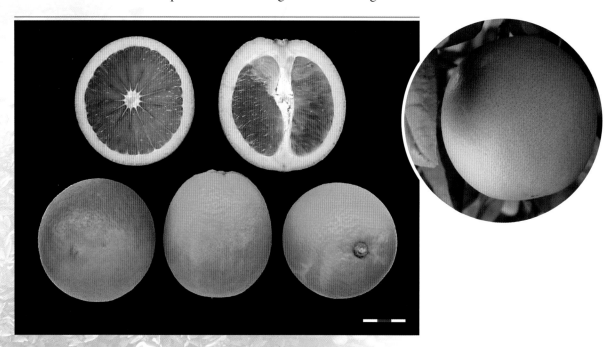

3. 塔罗科血橙

来源与分布：意大利主栽品种。我国20世纪80年代引入，重庆、四川、湖北等地有少量栽培。

主要性状：树势较强，春梢叶片大小为8.7cm×4.7cm，卵圆形或长椭圆形。果实倒卵形至椭圆形，橙黄色，完熟时带紫色斑纹，单果重150～200g，果顶平，果蒂部有明显沟纹，皮较厚，稍粗。TSS 11.0%～12.0%，酸0.8%～1.0%。果肉含花青素，呈血红色，柔嫩多汁，化渣，甜酸适中，少核或无核。成熟期1～2月。该品种丰产，果实耐贮运。

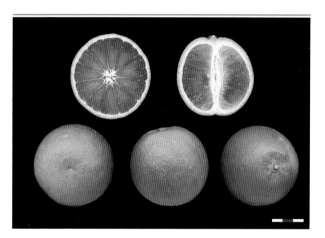

3. Tarrocco Blood Orange

Origin and Distribution: Tarrocco blood orange is the most popular variety in Italy. China introduced it in 1980s. It is limitedly cultivated in Chongqing, Sichuan and Hubei.

Main Characters: Tree comparatively vigorous; leaf size 8.7cm×4.7cm, ovate to long-elliptical. Fruit obovoid to ellipsoid, orange-yellow with purple strips after reaching complete ripening; weight 150~200g; apex flattened, base with prominent radial furrows; rind relatively thick and somewhat rough. Seedless or less seedy. TSS 11.0%~12.0%, acid content 0.8%~1.0%. Pulp containing anthocyanin, colored in blood-red, juicy, tender and melting with moderate sweet-sour flavor. Matures in January to February. The variety is productive, stores and ships well.

三、酸橙
Sour Oranges

1. 代代

来源与分布：原产我国，浙江、广东、广西有少量分布。

主要性状：树势健壮，树形稍直立，枝有短刺。叶椭圆形，先端渐尖，叶缘全缘或有微锯齿，翼叶明显。花中大，完全花，香味很浓。果实扁圆形，橙黄或橙红色，大小为7.5～7.9cm×6.8～7.2cm，果顶平，微凹。果蒂部有浅放射沟。果皮较粗糙，稍难剥皮。TSS 9.0%，酸3.0%～3.5%。囊壁厚，味酸。种子20～28粒/果，成熟期11月下旬。该品种适应性强，果实可在树上挂留2～3个月。可作香料、花茶原料，也可作观赏树木。

1. Daidai

Origin and Distribution: Daidai, originated in China, is limitedly cultivated in Zhejiang, Guangdong, and Guangxi.

Main Characters: Tree vigorous, somewhat erect, with short spines. Leaf elliptical, apex acuminate, margin entire or slightly dentate with pronounced petiole wing. Complete flower, medium-large, very fragrant. Fruit obloid, orange-yellow to orange-red, 7.5~7.9cm×6.8~7.2cm in size; apex truncate or slightly depressed, base with shallow radial furrows; rind somewhat rough, difficult to peel. TSS 9.0%, acid 3.0%~3.5%; segment membrane thick, taste sour. 20~28 seeds per fruit. Matures in late November. The variety is quite adaptable. Fruit could hold on tree for about 2~3 months. It can be used as flavor material of spices and tea, and also as ornamental plant.

2. 枸头橙

来源与分布：原产我国浙江，主产浙江等地。

主要性状：树势健壮，树冠圆头形，稍直立，枝条粗壮，有坚硬针刺。叶长椭圆形，先端钝圆，基部楔形，叶缘近于全缘，翼叶狭长。花有单花和花序花。果实圆形或扁圆形，淡黄色，大小为5.5~6.0cm×5.0~5.5cm，表皮油胞较稀，较粗糙。种子18~44粒/果，多胚。果肉味酸，有臭味。成熟期11月中下旬。该品种根群发达，耐盐碱能力较强，适应性广，生长快，寿命长，是浙江沿海宽皮柑橘的主要砧木之一。

2. Goutoucheng

Origin and Distribution: Goutoucheng, originated in Zhejiang, is cultivated mainly in Zhejiang Province.

Main Characters: Tree vigorous, spheroid, somewhat erect, branches sturdy with stiff spines. Leaf long-elliptical, apex obtuse to rounded, base cuniform, margin almost entire; petiole wing long and narrow. Solitary flower or inflorescence. Fruit spheroid to oblate, light yellow, 5.5~6.0cm×5.0~5.5cm in size; rind relatively rough, surface oil glands low in density and coarse. 18~44 seeds per fruit, polyembryonic. Pulp sour and smelly. Matures in mid to late November. The variety has well-developed root system, is widely adaptable with high tolerance to salty and alkaline soils, grows fast and survives for a long life. It is one of the major rootstocks of loose-skin mandarins in the coastal regions of Zhejiang Province.

3. 狮头橘

来源与分布：狮头橘又名虎头柑，原产广东，广东南澳岛栽培面积较大，春节作"发财柑"销售。

主要性状：树势强健，树冠自然圆头形，枝梢粗，略下垂。春梢叶片大小为7.5～10.0cm×4.4～5.6cm，阔椭圆形。花大，完全花。果实扁圆形，橙黄色，大小为11.3～12.0cm×7.7～8.5cm，果皮厚，果面粗糙。TSS 10.0%，酸1.5%～2.0%。种子15～25粒/果。成熟期1月中下旬。该品种粗生易种，果大，酸度大，多作药材或观赏用。以酸柚作砧木。

3. Shitouju

Origin and Distribution: Also named as Hutougan, this natural hybrid originated in Guangdong Province and is relatively extensively cultivated in Nanaodao of Guangdong. Its fruit is sold under the name of "Fortune Mandarin" during Spring Festival.

Main Characters: Tree vigorous, natural spheroid, branch somewhat thick, slightly drooping. Leaf 7.5~10.0cm×4.4~5.6cm, broad elliptical. Complete flower, large. Fruit obloid, orange-yellow, size 11.3~12.0cm×7.7~8.5cm. Rind thick, surface rough. TSS 10.0%, acid 1.5%~2.0%, 15~25 seeds per fruit. Matures in middle to late January. The variety is easy to grow. Fruit is large in size and high in acid content, and mainly cultivated for herb and ornamental uses. Sour pummelo is the common rootstock.

4. 朱栾

来源与分布：朱栾别名小红橙，原产浙江，浙江、广西、四川有少量分布。

主要性状：树势强健，树冠呈圆锥形，树姿稍直立，枝上多刺。叶椭圆形。花中大。果实馒头形，橙红色，大小为6.8～7.5cm×5.8～6.2cm，果肉味酸。TSS 9.0%，酸2.0%～2.5%。种子20～35粒/果。成熟期11月中下旬。该品种苗木前期生长快，根系发达，分布密集，须根多。浙江温州作瓯柑、温州蜜柑的砧木时，树较高大，结果期稍迟。

4. Zhuluan

Origin and Distribution: Zhuluan, also named as Small red orange, originated in Zhejiang Province. It has small-scale distribution in Zhejiang, Guangxi, Sichuan and probably other provinces.

Main Characters: Tree vigorous, conic, slightly erect, branches thorny. Leaf elliptical. Flower medium-sized. Fruit oblate, orange red, 6.8~7.5cm× 5.8~6.2cm in size, sour tastes. TSS 9.0%, acid 2.0%~2.5%, 20~35 seeds per fruit. Matures in middle to late November. Young seedling grows fast, developing a strong root system with densely distributed fibrous roots. In Wenzhou, Zhejiang, trees of Ougan and Satsuma mandarins on Zhuluan rootstock are relatively tall and slightly later to come into bearing.

四、柚/葡萄柚
Pummelos/Grapefruits

（一）柚 Pummelos

1. HB柚

来源与分布：HB柚又名江户文旦，原产美国，为文旦柚与葡萄柚天然有性杂交后代的实生变异。我国1990年从美国引入。经多年多点区试和示范，2001年通过湖北省农作物品种审定委员会认定。

主要性状：树势中等偏旺，树姿开张矮化，枝有短刺。叶片为卵圆形。果实扁圆形，橙黄色，单果重1 000～2 000g，果面光滑，油胞稀疏。TSS 10.0%～14.0%，酸0.9%～1.1%。成片种植籽少，混栽种子较多。果肉稍带粉红色，脆嫩化渣多汁。成熟期12月上旬。该品种丰产性好，单性结实力强，在无授粉树的情况下，果实种子少。无霜冻地区可留树贮至翌年3月，风味极佳。可用枳作砧木，与中间砧温州蜜柑、甜橙类均表现亲和。

1. HB Pummelo

Origin and Distribution: HB Pummelo, originated from America and also named as Jianghuwendan, is a seedling mutation among the offsprings of the cross between Wendanyou and grapefruit. It was introduced to China from America in 1990. After many-year and multi-place trial cultivation and demonstration, it was registered by Hubei Crop Cultivar Registration Committee in 2002.

Main Characters: Tree vigor medium to high, spreading and dwarfed, branches with short spines. Leaf ovate. Fruit oblate; orange-yellow, weight 1 000~2 000g, surface smooth, oil glands sparse. TSS 10.0%~14.0%, acid 0.9%~1.1%. Seeds few when single variety planted, relatively seedy when mix-planted. Pulp slightly colored in pink, crisp, tender, melting and juicy. Fruit matures in early December. The variety is quite productive, strongly parthenocarpic. Fruit could stores on tree until next March in frost-free regions with excelent flavor. Scion-rootstock compatibility is good when grafted on Trifoliate orange or interstocked with Satsuma mandarin or Sweet orange.

2. 安江香柚

来源与分布：本品种为湖南主要地方白柚良种之一，约有300年栽培历史，主产怀化地区。

主要性状：长势强旺，树冠阔圆头形，枝上有短刺。叶片大，长阔椭圆形，叶缘有不明显粗锯齿状，翼叶大，倒心形。果实长椭圆形或尖圆形，淡黄色，果面较光滑，单果重1 200～2 700g。TSS 13.8%左右，酸0.9%。汁胞乳白至淡米黄色，种子80～100粒/果。成熟期10月中下旬。果肉脆嫩化渣，富香气，无异味，丰产性好。果实耐贮性差些，贮至12月后易枯水粒化。以酸柚作砧木。

2. Anjiangxiangyou

Origin and Distribution: Anjiangxiangyou, cultivated for about 300 years, is one of the principal fine local white Pummelo varieties in Hunan Province and mainly grown in the district of Huaihua, Hunan.

Main Characters: Tree vigorous, broad spheroid, branches with short spines. Leaf large, long broad-elliptical, margin faintly crenate; petiole wing large, obcordate. Fruit long ellipsoid to pointed spheroid, light yellow, surface relatively smooth, weight 1 200~2 700g, 80~100 seeds per fruit. TSS about 13.8%, acid 0.9%. The color of juice vesicle is milk-white to light cream-yellow; Pulp crisp and tender, rich aroma, without unpleasant taste. The variety is quite productive and matures in mid to late November. Fruit stores slightly poorly, dehydrates and granulates as storage progresses to January. Common rootstock is Sour pummelo.

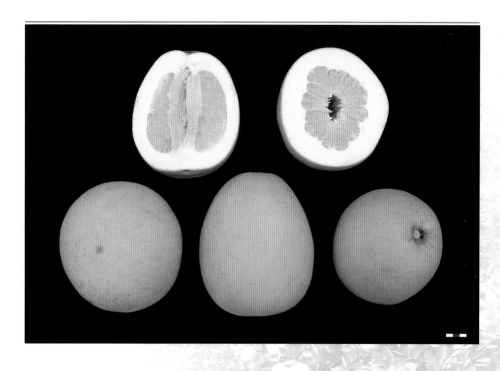

3. 处红柚

来源与分布：浙江省丽水林业科学研究所从实生柚中选出的早熟品种。2003年通过浙江省林木品种审定委员会认定。主产浙江丽水。

主要性状：树势中庸，树形开张，树冠自然圆头形，一年生枝密被茸毛。春梢叶片大小为8.0～15.0cm×5.0～10.0cm，卵形，翼叶大，倒心形。花大，完全花。果实梨形，橙黄色，大小为13.0～16.0cm×12.0～15.0cm，果面光滑，有光泽。果皮易剥离。海绵层淡红，囊壁薄而红，果肉深红色。TSS 12.0%，酸1.33%。少籽或无籽。果肉脆嫩，多汁，化渣，味甜酸适口。成熟期9月下旬。该品种丰产稳产性好，抗病性强，无裂果。以酸柚作砧木。

3. Chuhongyou

Origin and Distribution: Chuhongyou is an early ripening variety selected from a chance seedling of pummelos by Zhejiang Lishui Forest Science Institute, and was registered as a fine variety by Zhejiang Forest Cultivar Registration Committee in 2003.

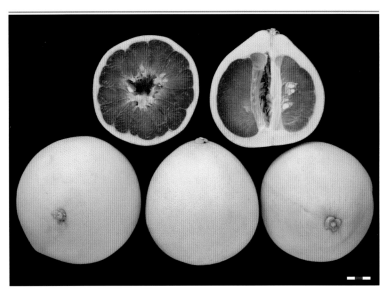

Main Characters: Tree medium-vigorous, spreading, natural spheroid, one-year old twigs densely pubescent. Leaf 8.0~15.0cm×5.0~10.0 cm in size, ovoid; petiole wing large, obcordate. Complete flower, large. Fruit pyriform, orange-yellow, 13.0~16.0cm×12.0~15.0cm in size, surface smooth and glossy; easily peeling; light red mesocarp, segment membrane is thin and red. Pulp is deep red, crisp, tender, juicy and melting with pleasantly sweet-sour tastes. TSS 12.0%, acid 1.33%. Few seeds or seedless. Matures in late September. The variety is resistant to diseases and has consistent and productive yield, without fruit splitting problem. Sour pummelo is often used as rootstock.

4. 垫江柚

来源与分布：原产重庆垫江，主产垫江等地。

主要性状：树势旺，树冠高大，枝叶茂密，小枝老熟后多光滑。春梢叶片大小为11.5cm×7.2cm，长椭圆形，叶缘有疏浅锯齿，翼叶较小。果实卵形，黄色，大小为15.2～16.6cm×15.0～17.2cm，果顶有不明显环纹，蒂部微凹有放射沟纹。TSS 11.0%～12.0%，酸0.8%～0.9%。种子60～80粒/果，多退化。成熟期11～12月。该品种果肉有红、白两种，熟期有早、中、晚类型。

4. Dianjiangyou

Origin and Distribution: Dianjiangyou, originated in Dianjiang, Chongqing, is mainly grown there.

Main Characters: Tree vigorous, tall; branches and leaves dense, twigs smooth when matures. Leaf 11.5cm×7.2cm in size, long elliptical, margin dentated sparsely and shallowly; petiole wing relatively small. Fruit ovoid, yellow, 15.2~16.6cm×15.0~17.2cm in size, apex with faint areole ring, base slightly depressed with radial furrows. TSS 11.0%~12.0%, acid 0.8%~0.9%; 60~80 seeds per fruit, mostly degenerated. Matures from November to September. The variety has two types of flesh colors: red and white, and three types of maturity: early, middle and late.

5. 东试早柚

来源与分布：云南西双版纳国营东风农场试验站从柚实生变异个体中选育出的早熟优良品种，又名水晶蜜柚。

主要性状：树势中强，树姿开张，树冠圆头形，枝条具浅刺。叶色深绿，叶片宽厚，叶缘浅波状，翼叶心形。花白色，总状花序或单生。果实倒卵形或锥形，果顶形状浅凹、放射沟纹明显，果基形状浅凹，果蒂周围放射沟纹明显程度中。成熟果实果皮黄色，白皮层白色，果面光滑度中等，油胞密而凸；单果重1 371g，大小为15.4cm×17.4cm，果形指数1.13，果皮厚1.4～1.9cm，易剥离，果心空；囊瓣肾形，13～15瓣，与囊壁易分离，果肉致密、白色；种子34～59粒/果，多退化；果味纯正，肉质嫩而化渣，甜酸适度，适口性好，TSS11.1%，酸0.8%，每100ml果汁维生素C含量47.5mg，可食率51.7%。结果母枝以春梢、秋梢为主；在云南瑞丽2月上旬开始出现春花蕾，2月下旬初花，3月中旬盛花，4月上旬终花，3月下旬出现二次生理落花高峰，4月下旬生理落果结束，坐果率为1.39%左右。东试早柚一般以枳、酸柚为砧木，以枳作砧木更加早熟、丰产、优质，最早8月上旬可成熟。适宜云南湿热区域，海拔500～900m，年均温20℃以上，年积温5 300～7 500℃，排水性好，pH 5.0～6.5，土层深厚的低丘、缓坡、平坝的柚适栽区。

5. Dongshizao Pummelo

Origin and Distribution: An excellent early-maturing variety, selected from the chance seedlings of Pummelo at the state-owned Dongfeng Farm Experimental Station in Xishuangbanna, Yunnan Province, also known as Crystal Honey Pummelo.

Main Characters: Tree medium vigorous, spreading, spheroid, branches with shallow thorns; dark green, broad and thick leaves with shallow undulate margin and heart-shaped petiole; white flowers, racemose or solitary. Fruit obovate or conical, apex slightly depressed with obvious radial furrows, base shallowly concave with moderately visible radial furrows. Mature fruit has yellow flavedo and white albedo, medium smooth surface with dense and rough oil glands. Average fruit weight 1 371g, height 17.4cm, width 15.4cm, fruit shape index 1.13, easy-to-peel pericarp thickness 1.4~1.9cm and fruit center hollow. Each fruit has 13~15 kidney-shaped sections easily separated from the segment membrane. with white compact pulp and 34~59 mostly degenerated seeds. Pulp tastes pure, tender, moderately sweet and sour, melting. TSS 11.1%, TA 0.8%, Vitamin C 47.5mg/100ml and edible portion 51.7%. Mainly spring or autumn shoots bear fruits. In Ruili, Yunnan, flowering begins in early February, blossom in mid-March, and ends in early April. The peak of physiological flower drop appears twice in late March, and physiological fruit drop ends in late April, with a fruit setting rate of about 1.39%. Generally, Trifoliate Orange or Sour Pummelo is used as rootstock for Dongshizao. Trifoliate Orange is preferred as the fruits can mature in early August with high yield and quality. Humid and hot region of Yunnan Province, with altitude 500 ~ 900m, annual average temperature above 20℃, annual accumulated temperature 5 300 ~ 7 500℃; on deep soil in the low hills, gentle slopes or flat dam regions with good drainage and pH 5.0 ~ 6.5, that suits for Pummelo cultivation.

6. 度尾文旦柚

来源与分布：度尾文旦柚又名仙游无核柚，原产福建省仙游县度尾镇，主产福建省。

主要性状：树冠半圆，自然开张。春梢叶片大小为16.0～17.0cm×6.0～7.5cm，椭圆形。花大，完全花，自交不亲和。果实扁圆形或高扁圆形，果基部稍尖，淡黄色，大小为16.0cm×14.8cm。果皮稍易剥离，中果皮淡粉红色或白色。TSS 12.2%，酸0.9%。单性结实力较强，种子败育或无种子。成熟期10月上旬。该品种果肉细嫩，甜酸适中，品质优。但果顶开裂较严重。以酸柚作砧木。

6. Duweiwendanyou

Origin and Distribution: Duweiwendanyou, also named as Xianyouwuheyou (seedless), originated in Weidu, Xianyou County, Fujian Province and is mainly grown in Fujian.

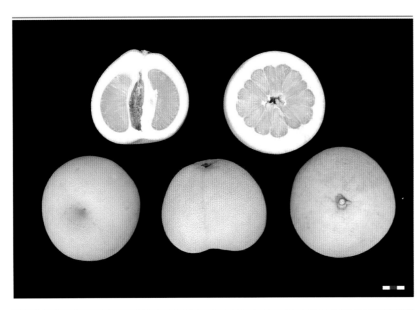

Main Characters: Tree semi-spheroid, naturally spreading. Leaf elliptical, 16.0~17.0cm×6.0~7.5cm in size. Complete flower, large, self-incompatible. Fruit oblate to high-oblate, base slightly protruded, light yellow, 16.0cm×14.8cm in size. somewhat easy peeling. Mesocarp colored in light pink or white. Fruit quality is good with 12.2% of TSS, 0.9% of acid and pulp tender with moderate sweet-sour flavor. The variety is strongly parthenocarpic with abortive seeds or seedless. However, fruit apex is prone to splitting. Matures in early October. Its common rootstocks is Sour pummelo.

7. 翡翠柚

来源与分布：浙江省丽水林业科学研究所从实生柚中选出，2002年通过浙江省农作物品种审定委员会审定。主产浙江丽水。

主要性状：树势中等，树冠自然圆头形。春梢叶片大小为8.0～15.0cm×4.0～7.0cm，翼叶大，花大，完全花。果实近倒卵形，淡绿色，大小为12.0～15.0cm×12.0～15.0cm，果顶平，微凹。果皮薄，较易剥离。TSS 10.0%～12.0%，酸0.3%。种子0～100粒/果。果肉淡绿色，味甜多汁，品质优。成熟期11月上旬，耐贮藏。该品种适应性广，抗性强。砧木用枳，用中熟温州蜜柑作中间砧，亲和力好。

7. Feicuiyou

Origin and Distribution: Feicuiyou, selected from a chance seedling by Zhejiang Lishui Forest Science Institute and registered by Zhejiang Crop Cultivar Registration Committee in 2002, is mainly grown in Lishui, Zhejiang Province.

Main Characters: Tree medium-vigorous, natural spheroid. Leaf 8.0~15.0cm×4.0~7.0cm in size, petiole wing large. Complete flower, large. Fruit near obovoid, light green, 12.0~15.0cm×12.0~15.0cm in size; apex truncate, slightly depressed; rind thin and relatively easy to peel. TSS 10.0%~12.0%, acid 0.3%, 0~100 seeds per fruit. Pulp light green, sweet and juicy. Fruit is of fine quality and stores well. The variety is widely adaptable and highly resistant. Fruit matures in early November. Scion-rootstock compatibility is good when grafted on Trifoliate orange rootstock or interstocked with Satsuma mandarin.

8. 琯溪蜜柚

来源与分布：琯溪蜜柚原产福建省漳州市平和县，主产福建，我国柚的栽培省份均有引种。

主要性状：生长势强，树冠圆头形或半圆形，枝叶稠密，内膛结果为主。春梢叶片大小为13.5～14.6cm×5.5～6.3cm，长卵圆形，叶缘锯齿浅，翼叶大，心脏形。花大，自交不亲和，单性结实能力强，不需人工授粉。果实倒卵形或梨形，果面光滑，中秋采收淡黄绿色，大小为16.9cm×15.7cm，果顶平，中心微凹且有明显印圈，成熟时金黄色，果皮稍易剥离。TSS 9.0%～12.0%，酸0.6%～1.0%。无籽，果肉甜，微酸。成熟期10月至11月上中旬。该品种是我国柚类的主栽品种，名柚之一。早结丰产，早熟，可中秋上市。贮藏性不及沙田柚，容易出现粒化。以酸柚作砧木。福建省农业科学院果树研究所已选出红肉蜜柚。

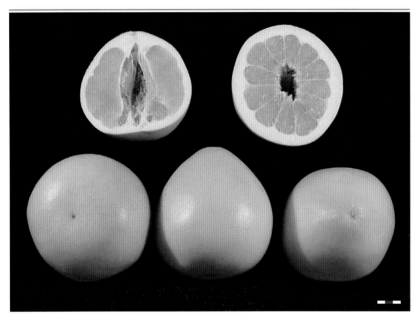

8. Guanximiyou

Origin and Distribution: Guanximiyou, originated in Heping County, Zhangzhou, Fujian Province, is mainly grown in Fujian and introduced to other pummelo producing provinces.

Main Characters: Tree vigorous, spheroid to semi-spheroid, branches dense. It mainly bears fruit inside the canopy. Leaf 13.5~14.6cm×5.5~6.3cm in size, long-ovate, margin shallowly dentate; petiole wing large, cordate. Large flower, self-incompatible, strongly parthenocarpic, requires no artificial pollination. Fruit obovoid to pyriform, smooth, light green-yellow when picked around the mid-autumn festival and golden yellow at maturity; 16.9cm×15.7cm in size; apex truncate or slightly depressed with conspicuous areole ring; slightly easy to peel. TSS 9.0%~12.0%, acid 0.6%~1.0%; seedless. Pulp is sweet, slightly acidic. Maturity period ranges from October to mid November. The variety is early in fruiting and maturing, very productive and becomes one of the major and famous pummelos in China. It can become marketable during the mid-autumn festival. Fruit storing quality is not as well as Shatianyou since the juice vesicle is prone to granulation. Common rootstock is Sour pummelo. Hongroumiyou has been selected out by Institute of Fruit Tree Research, Fujian Academy of Agricultural Sciences.

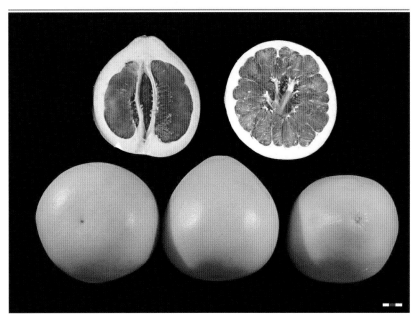

红肉蜜柚 Red flesh Guanximiyou

9. 桂柚1号

来源与分布：桂柚1号为沙田柚优良变异，2005年在广西恭城县平安乡发现，2010年通过广西农作物品种审定委员会审定。

主要性状：树势强，树冠较开张，萌芽力和成枝力强。在桂林，4月初至4月中旬开花，花期5~12d。结果母枝以春梢为主。自花结实，自然自花坐果率1.7%~9.7%。果实梨形，黄色，油胞大、明显，果顶微凹、印圈明显，单果重962.8~1 150.3g，大小为13.1~17.0cm×13.5~19.1cm，果形指数1.06~1.20，果皮厚1.61~2.10cm，囊瓣12~15瓣，种子91.3~155.7粒/果，可食率40.6%~47.8%，酸0.2%~0.4%、TSS10.4%~20.0%，全糖8.1%~11.1%，每100ml果汁维生素C含量72.5~109.6mg，风味甜、化渣，品质优良，耐贮性强。丰产稳产，在广西成熟期10月下旬至11月上中旬。适宜在沙田柚种植区推广。

9. Guiyou No.1 Pummelo

Origin and Distribution: Guiyou NO.1 is a variation of Shatianyou, found in Ping'an, Gongcheng County, Guangxi in 2005, and got registration in 2010.

Main Characters: Trees vigorous, crown open, budding and branching capacity strong. Flowering period 5 to 12 days, from ealry to mid-April in Guilin. Fruiting twigs are mainly spring shoots. Self-compatible with 1.7%~9.7% fruit setting rate. Fruit pyriform, rind yellow, oil glands large and immersed, apex slightly depressed with conspicuous areole ring. Fruit weight 962.8~1 150.3g, size in 13.1~17.0cm×13.5~19.1cm, fruit shape index 1.06~1.20, peel thickness 1.61~2.10 cm, segments 12~15, seeds 91.3~155.7 per fruit, edible portion 40.6%~47.8%, TSS 10.4%~20.0%, total sugar 8.1%~11.1%, TA 0.2%~0.4%, Vitamin C 72.5~109.6mg/100ml juice. Taste sweet, melting, quality excellent. Storage durable. Stably productive. Maturity in late October to mid November in Guangxi. Suitable for Shatianyou cultivation area.

10. 红宝石柚

来源与分布：红宝石柚又名泰国红肉柚、暹罗红宝石蜜柚，原产泰国南部洛坤府。我国2016年从泰国引进，广西、云南、广东、福建等柚的栽培省份均有引种。

主要性状：树势强健，树冠自然圆头形或半圆头形，较开张。春梢叶片阔椭圆形，叶缘锯齿浅，翼叶大，心脏形；春梢和夏梢为翌年主要结果母枝，嫩梢密被白色茸毛。花大，以总状花序为主，自交不亲和，单性结实能力强，不需异花授粉，可一年多次开花。果实梨形，有明显矮颈，果皮黄绿色，单果重1 121g，果顶平或微凹，无印圈，果皮薄，厚度约1.3cm，易剥离，幼果和成熟果表皮密被白色茸毛，无核，海绵层浅红色，囊壁薄而红。果肉红色，细嫩化渣，甜酸适口，果汁丰富，TSS9.0%～11.0%，酸0.3%～0.5%，每100ml果汁维生素C含量51.4～63.4mg。春季花果在桂南地区9月中下旬成熟，在桂中地区10月中下旬成熟。早结性较好，种植后第二年可结果。年均温22℃以上，积温高的区域为适栽区。

10. Hongbaoshi Pummelo (Ruby Pummelo)

Origin and Distribution: Hongbaoshi (Ruby) pummelo, also known as Thailand red-flesh pummelo, was previously produced in southern Thailand and introduced to Yunan, Guanxi, Guangdong and Fujian provinces China in 2016.

Main Characters: Trees very vigorous, spheroid to semi-spheroid, moderately spreading; Leaf broad elliptical, margin shallowly dentate, petiole wing large and cordate; spring and summer shoots are the main fruiting twigs for the following year. Young twigs densely pubescent. Flowers large, mostly racemose, self-incompatible, strongly parthenocarpic, no cross-pollination needed, bloom multiple times annually. Fruit pyriform, short-necked, rind yellow-green, average weight 1 121g, apex truncate or slightly depressed. Areole ring absent. Rind as thin as 1.3cm, somewhat easy to peel; surface of young and mature fruit pubescent, seedless; Albedo pink, segment membrane thin and red; Pulp red, tender, juicy and melting with TSS 9.0%~11.0%, TA 0.3%~0.5% and Vitamin C 51.4~63.4mg/100ml. Fruits mature in mid-to-late September in southern Guangxi, mid-to-late October in central Guangxi. Precocious, fruit bearing from the second year of cultivation. It is suitable for growing in areas where the average annual temperature is above 22℃, with high accumulated temperature and high heat.

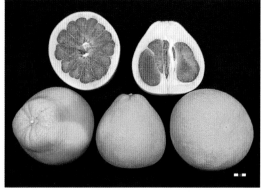

11. 华柚2号

来源与分布：华柚2号是国庆1号温州蜜柑愈伤组织原生质体与华柚1号（HB柚）叶肉原生质体融合创制的二倍体雄性不育胞质杂种，由华中农业大学2004年创制，2015年获得植物新品种权。

主要性状：树势中等，树姿开张。叶形卵圆，叶尖钝尖，叶基圆形。花瓣短而退化，雄蕊败育，雌蕊发育正常。果实大，扁圆形，果皮橙黄色，果面较光滑，油胞少，单果重1 232g，果形指数0.86，果顶浅凹，无印圈。果皮较难剥离，白皮层颜色浅粉红，中心柱中等大，果心空，囊瓣肾形，10~15瓣。果肉粉红色，囊壁薄，化渣性好，多汁，风味浓，有香气。与雄性不育品种混栽或隔离种植时果实无核，可食率69.0%~80.1%，TSS11.5%~13.3%，酸0.7%~1.2%，固酸比10.4~16.3，每100 ml果汁维生素C含量42.4mg。在湖北武汉地区，果实11~12月成熟，五年生嫁接树单株产量约50 kg，无开裂果。耐贮藏，抗逆性和适应性较强，耐寒性中等。适宜在中国南方柚种植区推广。

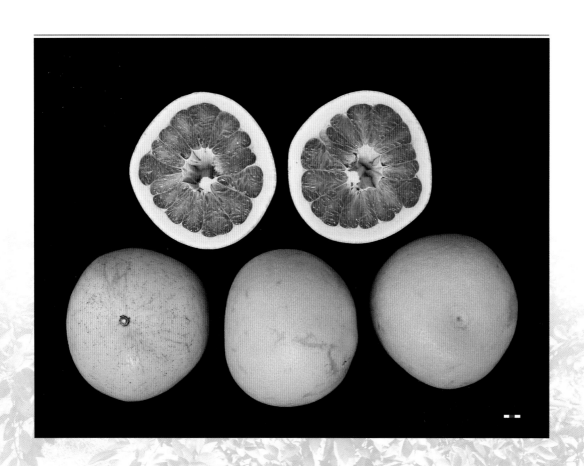

11. Huayou No.2 Pummelo

Origin and Distribution: Huayou No.2 pummelo is a male-sterile somatic hybrid produced by protoplast fusion with Guoqing No.1 Satsuma Mandarin and Huayou No.1 (Hirado Buntan pummelo). It was generated by Huazhong Agricultural University in 2004 and got PVP of the Ministry of Agriculture in 2015.

Main Characters: Tree medium-vigorous and spreading. Leaves narrow ovate, apex blunt tip, base round, flower petals aberrant, and stamens sterile, pistils normal. Fruit size big, oblate, surface smooth, orange-yellow, with few oil glands; Average fruit weight 1 232 g, fruit shape index 0.86, apex slightly depressed, without areole ring. Rind moderately difficult to peel, albeto light pink, axis medium and hollow, segments 10 to 15, kidney-shaped. Pulp pink, juicy and melting with rich flavor and aroma. Seedless when cultivated with other male-sterile cultivars or separately from other fertile cultivars, edible portion 69.0%~80.1%, TSS 11.5%~13.3%, TA 0.7%~1.2%, Vitamin C 42.4 mg/100ml. In Wuhan, the fruit ripening period is during November and December. The average yield of 5-year grafted tree can reach 50 kg without fruit cracking. Storage durable, resistance and adaptability strong, cold-hardiness medium. It is suitable for pummelo cultivation areas in southern China.

12. 金兰柚

来源与分布：原产我国，广东紫金、江西等地有零星栽培。

主要性状：树势中等，树冠扁圆头形，枝梢粗。春梢叶大小为12.0～14.0cm×5.0～5.5cm，椭圆形，翼叶中等大。花大，完全花。果实高椭圆形，橙黄色，大小为17.5～19.0cm×18.6～20.0cm，果顶平，果蒂微凹，有不明显4～5条放射沟纹。果皮中等厚，有光泽，稍易剥离。TSS 10.5%～12.0%，酸0.5%～0.6%。种子100～120粒/果。成熟期11月上中旬。果肉脆嫩，味甜少酸，成熟后果实香味浓。丰产性好。以酸柚作砧木。

12. Jinlanyou

Origin and Distribution: Jinlanyou originated in China and now its cultivation scattered in Zijin, Guangdong Province and Jiangxi Province.

Main Characters: Tree medium-vigorous, crown oblate, branches thick. Leaf 12.0~14.0cm×5.0~5.5cm in size, elliptical, petiole wing medium-large. Complete flower, large. Fruit long elliptical, orange-yellow, 17.5~19.0cm×18.6~20.0cm in size, 100~120 seeds per fruit; apex flatten; base slightly concave, striped with 4~5 faint radial furrows. Rind medium-thick, glossy and easily peeling. TSS 10.5%~12.0%, acid 0.5%~0.6%. Pulp crisp and tender with sweet and less sour taste. Fruit matures in early to middle November. Fruit is quite fragrant at maturity. The variety is quite productive. Sour pummelo is used as rootstock.

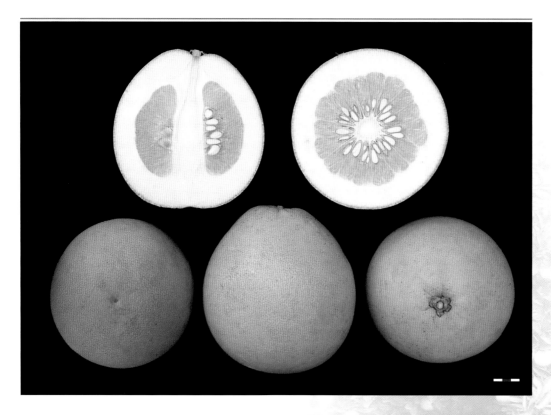

13. 金香柚

来源与分布：原产湖南省慈利县，主产湖南。

主要性状：树冠高大，长圆头形，枝条硬而直立，有刺。叶片长椭圆形，叶缘波状锯齿，翼叶倒锥形，花为总状花或单生。果实椭圆柱形，橙黄色，大小为10.5cm×13.5cm，果顶平钝，有放射沟纹。果皮较薄，较易剥离。TSS 13.0%～15.0%，酸0.4%。成熟期9月下旬至10月中旬。果肉米黄色，柔软多汁，味甜而浓香，品质上等，耐贮藏。以酸橙或酸柚作砧木。

13. Jinxiangyou

Origin and Distribution: Jinxiangyou, originated in Cili County, Hunan Province, is mainly grown in Hunan.

Main Characters: Tree tall, long spheroid, branches hard and erect with spines. Leaf long elliptical, margin sinuate; petiole wing obdeltate. Solitary flower or raceme. Fruit elliptical-columniform, orange yellow, 10.5cm×13.5cm in size; apex obtusely truncate, with shallow radial furrows; rind relatively thin, comparatively easy to peel. TSS 13.0%~15.0%, acid 0.4%. Maturity period ranges from late September to mid-October. The pulp of the variety is cream-colored, soft, juicy, sweet, and very fragrant. Fruit has superior quality and stores well. Sour orange or Sour pummelo is often used as rootstock.

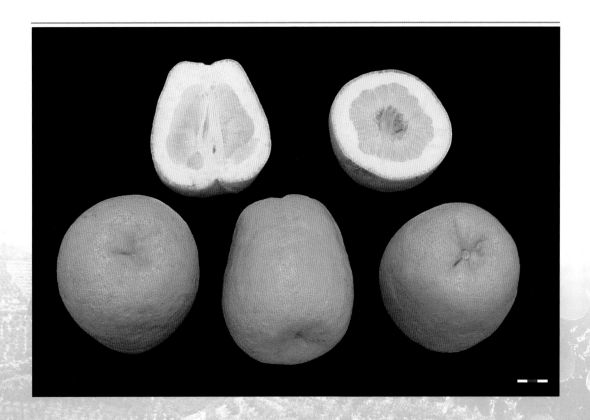

14. 橘红

来源与分布：原产广东省化州市，在化州市建有橘红生产基地，用于药材原料。

主要性状：树势中等，树冠圆头形，枝梢较密。春梢叶片大小为12.0～13.0cm×6.0～6.6cm，呈卵圆形，叶翼大。花稍大，完全花。果实近圆形，淡黄色，果皮密生灰白色茸毛。果汁少，味酸稍带苦，不堪食用。幼果是止咳化痰的特种药材。砧木用酸柚。

14. Juhong

Origin and Distribution: Juhong originated from Huazhou, Guangdong Province. The production base of Juhong has been set up in Huazhou for the material of Chinese medicine.

Main Characters: Tree middle vigorous. Crown spheroid, branches dense. Leaf size 12.0~13.0cm×6.0~6.6cm, ovate. Leaf wing large. Complete flower, relatively bigger. Fruit shape is round. The color of rind is yellow at maturity with many small whitish hairs. Flesh is less juicy, sour and bitter. Unmature fruit is used as a Chinese medicine for treating cough and expectoration. The main rootstock is Sour pummelo.

15. 梁平柚

来源与分布：原产重庆市梁平县，主产重庆市。

主要性状：长势中等，树冠矮小，较开张，枝条粗糙，多下垂。春梢叶片大小为11.4cm×6.8cm。果实高扁圆形，橙黄色，大小为14.7cm×11.0cm，果顶平凹，基部平。果皮薄，光滑，有香味，稍易剥离。TSS 11.0%～12.0%，酸0.3%。成熟期10月下旬。该品种适应性强，丰产稳产，果肉细嫩多汁，化渣，味浓甜，有点苦麻味。较耐贮藏。以酸柚作砧木。

15. Liangpingyou

Origin and Distribution: Liangpingyou, originated in Liangping County, Chongqing, is mainly grown in Chongqing.

Main Characters: Tree medium-vigorous, dwarfed and small, relatively spreading; branches rough, mostly drooping. Leaf 11.4cm×6.8cm in size. Fruit high oblate, orange-yellow, 14.7cm×11.0cm in size, apex truncate to depressed, base truncate. Rind is thin, smooth, fragrant and somewhat easy peeling. TSS 11.0%~12.0%, acid 0.3%. Matures in late October. Pulp is tender, juicy, melting and rich sweet with slight bitterness. The variety is quite adaptable with consistent and high yield. Fruits stores relatively well. Sour pummelo is often used as rootstock.

16. 坪山柚

来源与分布：原产福建省华安县新圩镇坪山村，是我国名柚之一，宋时已广为栽培，约有600年历史。主产福建省，广东、浙江有少量栽培。

主要性状：树姿开张，枝条略披垂，树冠圆头形。春梢叶片大小为16.0cm×7.0cm，椭圆形。花较大，完全花。果实倒卵形，大小为15.0～17.0cm×17.0～19.0cm，果蒂部圆、略小、有凹入，常歪斜一边，果顶平。中果皮及囊衣浅红色。TSS 11.0%～12.5%，酸0.6%。成熟期9月下旬至10月上旬。果肉脆，味清甜，质优。种子较多，每果100粒左右。以酸柚作砧木。

16. Pingshanyou

Origin and Distribution: Pingshanyou, originated in Pingshan village, Huaan County, Fujian Province, is a famous pummelo variety in China. It was widely cultivated in the Song Dynasty, which was 600 years ago. Now is grown mainly in Fujian Province and limitedly in Guangdong and Zhejiang.

Main Characters: Tree spreading, spheroid, branches slightly dropping. Leaf 16.0cm×7.0cm in size, elliptical. Complete flower, relatively large. Fruit obovoid, 15.0~17.0cm×17.0~19.0cm in size; base rounded or slightly depressed, commonly oblique; apex truncate; mesocarp and segment membrane colored in light red; TSS 11.0%~12.5%, acid 0.6%. Pulp is crisp and fresh sweet. Fruit quality is good though seedy, about 100 seeds per fruit. Maturity period ranges from late September to early October. Sour pummelo is its common rootstock.

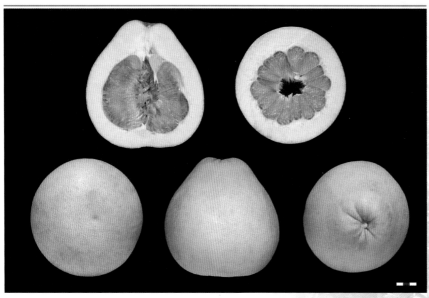

17. 强德勒柚

来源与分布：1979年从美国引进。现浙江、重庆、福建、湖北、广东等地有少量栽培。

主要性状：树势强健，树冠高大，枝梢较直立，结果后开张。春梢叶片大小为15.0cm×6.0cm，长椭圆形。单性结实力较强。果实近球形或倒阔卵形，黄色，大小为14.9cm×13.3cm，果面光滑，果皮稍薄，难剥离。TSS 11.0%，酸0.9%，种子70～90粒/果。果肉红色，脆嫩，果汁中等，酸甜可口，无异味，品质优。成熟期11月中旬。该品种早结丰产，耐贮藏。砧木以枳、枳橙、枳柚、酸柚均可。

17. Chandler Pummelo

Origin and Distribution: Chandler pummelo, introduced from America in 1979, is now cultivated to some extent in Zhejiang, Chongqing, Fujian, Hubei and Guangdong etc.

Main Characters: Tree vigorous, tall, branches somewhat erect, spreading after fruiting. Leaf 15.0cm×6.0cm in size, long elliptical. Strongly parthenocarpic. Fruit nearly spheroid to broad obovoid, yellow, 14.9cm×13.3cm in size, surface smooth; pericarp slightly thin, difficult to peel; TSS 11.0%, acid 0.9%, 70~90 seeds per fruit. Pulp is red, crisp, tender and medium-juicy with pleasant sour-sweet taste without bitterness. The fruit matures in mid-November with good quality. The variety is early fruiting, productive, and stores well. Trifoliate orange, Citrange, Citrumelo, or Sour pummelo can be used as its rootstock.

18. 三红蜜柚

来源与分布：三红蜜柚为红肉蜜柚芽变，2004年在福建省平和县小溪镇红肉蜜柚果园发现，2013年通过福建省农作物品种审定委员会认定。

主要性状：树势强健，萌芽力中等，成枝力强，幼树较直立，成年树半开张，树冠半圆头形，以春梢为主要结果母枝。单果重1 400g，果皮在适当遮阳或套袋后呈淡粉红色，海绵层粉红色，果肉玫瑰红色、含番茄红素，无籽，汁胞柔软多汁，清甜微酸，TSS10.0%，酸0.5%，每100ml果汁维生素C含量36.5mg。平和县低海拔区域完全成熟在10月中下旬。该品种丰产、稳产，适宜在年均温20℃以上、积温6 500℃以上区域栽培。

18. Sanhongmiyou

Origin and Distribution: Sanhongmiyou, a pink-flesh bud mutation. It was found in 2004 in Xiaoxi Town, Pinghe County of Fujian Province, and registered in Fujian in 2013.

Main Characters: Tree vigorous with medium branching capacity, upright in youth, and semi-spreading after fruiting, crown semi-spheroid. The fruiting twigs are mainly spring shoots. Fruit weight 1 400g on average, rind light pink when proper shading or bagging, albedo pink, pulp rosy-red and containing lycopene. Seedless, soft and juicy, fresh sweet, slightly acid. With TSS 10.0%, TA 0.5%, Vitamin C 36.5mg/100ml. Full maturity in mid-to-late October in the low altitude area of Pinghe County. Prolific and yield stable. It is suitable to be cultivated in the area where the annual average temperature is above 20℃ and the accumulated temperature is above 6 500℃.

19. 桑麻柚

来源与分布：珠江三角洲顺德一带农家品种，1761年《五山志林》已有记载。现广东省紫金县种植较多，博罗、顺德等地有分布。

主要性状：长势中等，树冠圆头形，枝条稍粗。春梢叶片大小为10.8～13.3cm×5.8～7.0cm，椭圆形，翼叶较小。花大，完全花。果实倒卵圆形，深黄色，大小为16.0～18.0cm×16.5～19.0cm，果蒂有放射沟纹，果顶平，果面微有凹凸，油胞较粗。果皮稍厚，稍易剥离。TSS 10.0%～12.0%，酸0.5%～0.6%，种子50～60粒/果。成熟期中秋节期间。果肉爽脆汁多，味甜，品质中上，早熟。以酸柚作砧木。

19. Sangmayou

Origin and Distribution: Sangmayou is a local garden pummelo variety in Shunde of Zhujiang Delta area, Guangdong. This ancient variety was recorded in "Wushanzhilin" in 1761. Now, it's cultivation mainly centers in Zijin County, Guangdong Province. There are some plantings in Boluo and Shunde, Guangdong.

Main Characters: Tree medium-vigorous, spheroid, branches somewhat thick. Leaf 10.8~13.3cm×5.8~7.0cm in size, elliptical; petiole wing relatively small. Complete flower, large. Fruit obovoid, deep yellow, 16.0~18.0cm×16.5~19.0cm in size; base low collared with radial furrows; apex flatten, surface slightly bumpy, oil glands comparatively large, pericarp somewhat thick, easily peeling. TSS 10.0%~12.0%, acid 0.5%~0.6%; 50~60 seeds per fruit. Pulp is fresh, crisp and juicy with sweet taste. Fruit is of medium to high quality and early maturing at around Mid-Autumn festival. Sour pummelo is its common rootstock.

20. 沙田柚

来源与分布：沙田柚原产广西容县沙田，现主产广西、广东、湖南、重庆等地。

主要性状：树势强，树冠圆头形或塔形，枝梢稍粗，内膛结果为主。春梢叶片大小为10.0～12.0cm×6.0～7.0cm，椭圆形，叶缘波浪形，锯齿较浅，翼叶大。花大，完全花，自交不亲和，单性结实能力弱，需人工授粉结实率才高。果实梨形或葫芦形，橙黄色，大小为13.5～15.0cm×16.5～17.5cm，果顶部平或微凹，有不整齐的印环，环内稍突出，果蒂有长颈和短颈两种，短颈品质较好。果皮中等厚，稍难剥离。TSS 12.8%～16.0%，酸0.4%～0.5%，果肉脆嫩浓甜。种子60～120粒/果。成熟期11月中下旬。该品种是我国柚类主栽品种，名柚之一。果实耐贮运，以酸柚作砧木。

20. Shatianyou

Origin and Distribution: Shatianyou, originated in Shatian, Rongxian, Guangxi, is mainly cultivated in Guangxi, Guangdong, Hunan, Chongqing.

Main Characters: Tree vigorous, spheroid to tower-like, branches somewhat thick. It bears fruit inside canopy. Leaf 10.0~12.0cm×6.0~7.0cm in size, elliptical, margin sinuate, shallowly dentate, with large petiole wing. Complete flower, large, self-incompatible, weakly parthenocarpic. Artificial pollination is required to achieve high yield. Fruit pyriform or cucurbit, orange-yellow, 13.5~15.0cm×16.5~17.5cm in size; apex truncate or slightly depressed with irregular areole ring, the inside of the ring slightly protruded; base long-necked or short-necked, the short-necked fruit has better quality. Rind medium-thick, somewhat difficult to peel. TSS 12.8%~16.0%, acid content 0.4%~0.5%. Pulp crisp and tender with rich sweet flavor. 60~120 seeds per fruit. Matures in middle to late November. Shatianyou is one of the major and the most famous pummelo varieties in China. Fruit stores and ships well. Sour pummelo is commonly used as its rootstock.

21. 四季抛

来源与分布：四季抛又名四季柚，原产浙江省苍南，从土柚的实生变异中选出。主产浙江、福建、广东、广西等地有栽培。

主要性状：树势中等，树冠圆形或半圆形，枝条节间短，多茸毛，春梢叶片大小为10.8cm×5.9cm，阔椭圆形。叶缘无锯齿或锯齿不明显。花为总状花序或单花。果实倒卵形，黄色，大小为9.5～12.5cm×8.3～11.0cm，顶部广圆。果皮薄，稍难剥离。TSS 10.0%～13%，酸0.7%～0.8%。无核，果肉味酸甜，品质中等。成熟期11月上中旬。该品种适应性强。以酸柚作砧木。

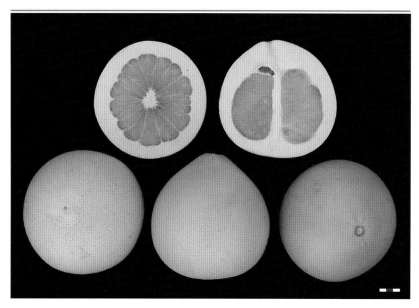

21. Sijipao

Origin and Distribution: Sijipao, also named as Sijiyou, originated in Cangnan, Zhejiang Province, as a chance seedling of local Pummelo and is mainly grown in Zhejiang and Fujian. It was also cultivated to some extent in Guangdong, Guangxi and etc.

Main Characters: Tree medium-vigorous, spheroid to semi-spheroid; branches with short internode distance, densely pubescent. Leaf 10.8cm×5.9cm in size, broad elliptical, margin entire or faintly serrate. Solitary flower or raceme. Fruit obovoid, yellow, 9.5~12.5cm×8.3~11.0cm in size, apex broad rounded, seedless. Rind thin and somewhat difficult to peel. TSS 10.0%~13.0%, acid 0.7%~0.8%, sour-sweet taste. Matures in early to middle November. The variety has medium fruit quality and is quite adaptable and commonly grafted on Sour pummelo.

22. 酸柚

来源与分布：原产我国，是我国古老品种。全国柚产区均有零星栽培。

主要性状：树冠圆头形，枝条具刺。叶片大小10.0～11.0cm×5.2～6.2cm，椭圆形，叶缘锯齿浅，翼叶大。花大，完全花。果实扁圆形或圆形，青黄或橙黄色，大小为13.0～14.0cm×11.5～12.6cm，果顶平，或微凹，果蒂浑圆，果皮稍易剥离。TSS 11.5%～12.5%，酸1.2%～1.5%，种子90～130粒/果。果肉有红、白等色，酸带苦，不堪食用，主要作砧木。

22. Sour Pummelo

Origin and Distribution: Sour pummelo, originated in China, is an ancient Chinese variety and cultivated limitedly in almost all Pummelo producing areas in China.

Main Characters: Tree spheroid, branches with spines. Leaf 10.0~11.0cm×5.2~6.2cm in size, elliptical, margin shallowly dentate, petiole wing large. Complete flower, large. Fruit oblate to spheroid, green-yellow to orange-yellow, 13.0~14.0cm×11.5~12.6cm in size. Apex flatten, or slightly depressed. Base rounded. Relatively easy to peel. TSS 11.5%~12.5%, acid 1.2%~1.5%. 90~130 seeds per fruit. Pulp is red or white, sour and bitter, no edible value, commonly used as rootstock.

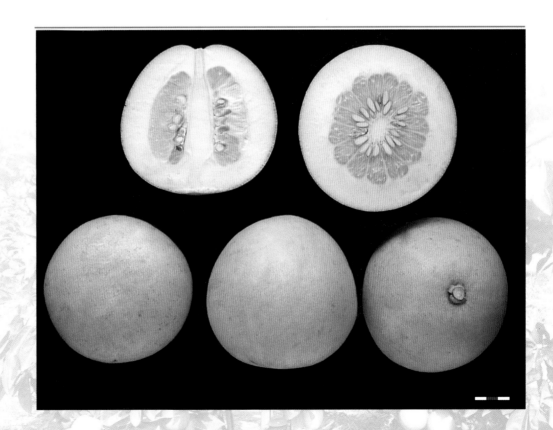

23. 晚白柚

来源与分布：原产我国台湾，台湾、福建、重庆、四川等地有栽培。

主要性状：树势强，树冠圆头形，较矮小。枝披垂，密被茸毛。叶片大小为12.9cm×6.6cm，长椭圆形，全缘，叶缘披短毛，翼叶大，心脏形。花大，自交不亲和。果实扁圆形或圆形，橙黄色，大小为14.0cm×13.0cm，果顶平，印圈不明显。TSS 12.0%，酸1.06%。汁胞白色或浅黄绿色，肉质细嫩，多汁，化渣，有香气，品质优。成熟期12月至翌年1月。该品种较早结丰产，果大。以酸柚为砧木。

23. Wanbaiyou

Origin and Distribution: Wanbaiyou, originated in Taiwan Province, is grown to some extent in Taiwan, Fujian, Chongqing, Sichuan.

Main Characters: Tree vigorous, spheroid, relatively dwarfed, branches drooping, densely pubescent. Leaf 12.9cm×6.6cm in size, long elliptical, margin entire, puberulent. Petiole wing large, cordate. Flower large, self-incompatible. Fruit oblate to spheroid, orange-yellow, 14.0cm×13.0cm in size. Apex flatten with faint areole ring. TSS 12.0%, acid 1.06%. The color of juice vesicle is white or light yellow-green. Flesh is tender, juicy, melting and fragrant. Fruit quality is fine and matures from December to next January. The variety is relatively early-fruiting and productive. Sour pummelo is used as its rootstock.

24. 玉环柚

来源与分布：玉环柚又名楚门文旦。原产浙江玉环县楚门，是福建文旦柚的实生变异，主产浙江省。

主要性状：树势健壮，树冠圆头形，较开张，枝粗壮，有小刺。春梢叶片大小为15.5cm×6.1cm，长椭圆形。花大，完全花，自交不亲和，单性结实力较强。果实有梨形、高扁圆形和扁圆形，大小为16.4cm×15.0m，顶端微凹，基部较窄，有浅沟纹，果皮稍易剥离。TSS 11.0%～12.0%，酸1.0%～1.2%。种子多数退化或无核。成熟期10月中下旬。该品种丰产性一般，果实扁圆类含糖量较高，风味浓，生产上栽培较多，但容易裂果。以酸柚或枸头橙作砧木。

24. Yuhuanyou

Origin and Distribution: Yuhuanyou, also named as Chumen-wendan and originated in Chumen, Yuhuan County, Zhejiang Province, is a chance seedling of Fujian Wendanyou and mainly grown in Zhejiang Province.

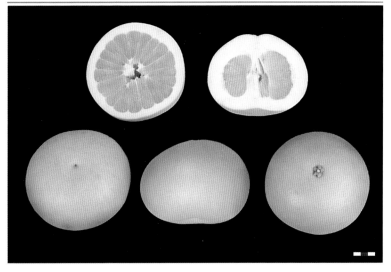

Main Characters: Tree vigorous, spheroid, somewhat spreading, branches sturdy with small spines. Leaf 15.5cm×6.1cm in size, long elliptical. Complete flower, large, self-incompatible, strongly parthenocarpic. Fruit pyriform, long-oblate or oblate, 16.4cm×15.0cm in size; apex slightly depressed; base relatively narrow, with shallow grooves; comparatively easy to peel. TSS 11.0%~12.0%, acid content 1.0%~1.2%; seeds degenerated or seedless; oblate fruit has higher sugar content and richer flavor, matures in middle or late October. The variety is medium productive and is widely cultivated although the fruit is prone to splitting. The commonly used rootstocks are Sour pummelo and Goutoucheng.

25. 早香柚

来源与分布：早香柚又名永嘉香抛。浙江省永嘉县从文旦实生柚选出，1990年定名为永嘉早香柚。主产浙江省。

主要性状：树势强健，树冠紧凑。枝较短，丛生状。叶片披针状椭圆形。果实梨形，橙黄色，大小为16.0cm×19.0cm，果实成熟后散发出浓郁的香气，贮藏愈久香气愈浓。TSS 10.5%～13.5%，酸0.8%～0.9%。果肉米黄色，晶莹透亮，脆嫩，多汁，甜酸适口，有清香，无苦麻味，品质优。少核，种子20～40粒/果（其中退化种子6～10粒），成熟期9月下旬。该品种丰产，早熟。以酸柚作砧木。

25. Zaoxiangyou

Origin and Distribution: Zaoxiangyou, also named as Yongjiaxiangpao, was selected from a chance seedling of Wendanyou by Yongjia County, Zhejiang Province, and is mainly grown in Zhejiang.

Main Characters: Tree vigorous, compact-topped; branches relatively short, clustered. Leaf lanceolate to elliptical. Fruit pyriform, orange-yellow, 16.0cm×19.0cm in size, with very fragrant aroma at maturity and even more fragrant after storage. TSS 10.5%~13.5%, acid 0.8%~0.9%. Pulp is cream-yellow, glittering, translucent, crisp, tender and juicy with pleasant sweet-sour taste and fresh aroma. Fruit quality is fine without any bitter sense. Seeds few, 20~40 seeds per fruit (of which 6~10 seeds degenerate). Matures in late September. The variety is productive and early in maturity. Sour pummelo is its common rootstock.

26. 早玉文旦

来源与分布：早玉文旦是玉环柚中选出的早熟品种，2017年通过浙江省林木良种认定。

主要性状：树势强健，树冠圆头形。果实高扁圆形，单果重1 250～2 000g，TSS12.5%，酸0.8%。果汁丰富，脆嫩爽口，品质优。特早熟，成熟期为9月10日，比普通玉环柚早40d左右。丰产优质，较易裂果，需要异花授粉，宜适度推广。

26. Zaoyuwendanyou Pummelo

Origin and Distribution: An early-season variety selected from Yuhuan pummelo and registered in Zhejiang Province in 2017.

Main Characters: Tree vigorous with a spheroid crown. Fruit long obloid, weight 1 250~2 000g, TSS 12.5% and TA 0.8%. Fruit is juicy, crisp, tender and tasty. It is of excellent quality and highly precocious. The mature period is around September 10th, which is about 40 days earlier than ordinary Yuhuan pummelo. Although this elite cultivar is highly productive, it needs cross-pollination and the fruit tends to crack. It should be promoted moderately.

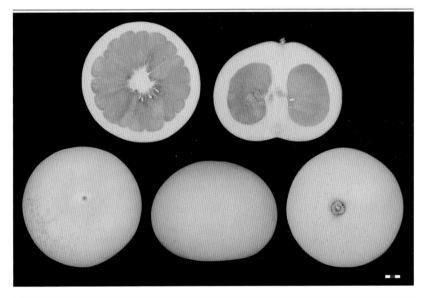

（二）葡萄柚　Grapefruits

1. 奥兰布兰科柚

来源与分布：美国以二倍体无酸柚和四倍体白肉葡萄柚杂交获得的三倍体柚品种。我国从美国、以色列引入。现湖北、四川、广东有少量种植。广东通过多年多点试验，2003年通过广东省农作物品种审定委员会认定。

主要性状：树形开张，树势强健。春梢叶片大小为10.4cm×7.1cm，卵圆形，叶缘锯齿浅，翼叶中大。果实扁圆形，黄绿色有光泽，大小为10.0～10.5cm×8.5～10.6cm，皮肉稍易分离。TSS 9.0%～11.0%，酸0.5%～0.6%。无核，果肉柔软、多汁、化渣。成熟期11月上中旬，气温稍高的广州9月上旬可上市。该品种丰产性较好。以枳橙或酸柚作砧木。

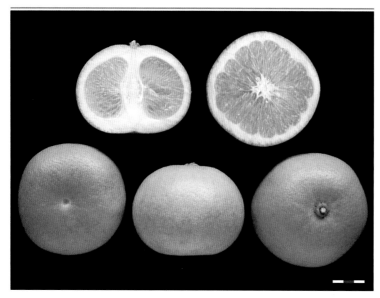

1. Oroblanco

Origin and Distribution: Oroblanco is a triploid pummelo variety derived from a cross between a diploid acidless pummelo and the tetraploid white-pulp grapefruit made in America. China introduced it from America and Israel. Currently it is limitedly cultivated in Hubei, Sichuan and Guangdong. It was registered in 2003 by Guangdong Crop Cultivar Registration Committee after multi-point trial cultivation for many years.

Main Characters: Tree spreading and vigorous. Leaf 10.4cm×7.1cm, ovate, margin shallowly dentate, petiole wing medium-large. Fruit oblate, yellow-green, glossy, 10.0~10.5cm×8.5~10.6cm in size, somewhat easy to peel, seedless. TSS 9.0%~11.0%, acid 0.5%~0.6%. Pulp is tender, juicy and melting. Fruit matures in early to mid-November and can be marketable in early September in Guangzhou, Guangdong Province, where the temperature is higher. The variety is relatively productive. Citrange or Sour pummelo are commonly used as its rootstock.

2. 菠萝香柚

来源与分布：原产湖南慈利县，是橙和柚的天然杂交种，主产湖南省。

主要性状：树冠圆头形，枝梢生长较密，略披垂。春梢叶片大小为8.4cm×4.1cm，披针形，叶缘浅圆锯齿形，翼叶较大。果实近球形，橙黄色，大小为9.4cm×8.3cm，果皮易剥离。TSS 11.0%～12.0%。果肉化渣多汁，甜酸适中，风味浓，有菠萝香味。成熟期11月中旬。该品种果形美观，耐贮藏。

2. Boluoxiangyou

Origin and Distribution: Boluoxiangyou, originated in Cili County, Hunan Province as a natural cross of orange and pummelo, is mainly grown in Hunan.

Main Characters: Tree spheroid, branches relatively dense, somewhat drooping. Leaf 8.4cm×4.1cm in size, lanceolate, margin shallowly crenate, petiole wing relatively large. Fruit nearly spheroid, orange-yellow, 9.4cm×8.3cm in size, easy peeling. TSS 11.0%~12.0%. Pulp is melting and juicy with moderate sweet-sour taste and rich flavor. Fruit matures in mid November with pineapple aroma and stores well.

3. 胡柚

来源与分布：原产浙江常山，是柚与甜橙的自然杂交种。主产浙江省。

主要性状：树势强，树冠圆头形，枝梢稍直立。果实梨形或扁球形，橙黄色，大小为9.6cm×4.2cm，果顶有明显或不明显的印圈，有粗皮和细皮之分，果皮易剥离。TSS 11.0%，酸1.0%～1.3%。种子10～40粒/果，少核种3～4粒/果，间有无核。果肉甜酸适度，略带苦味，风味浓爽可口，品质较优。成熟期11月中下旬。该品种丰产性好，较抗寒，果实耐贮藏。以枳作砧木。

3. Huyou

Origin and Distribution: Huyou, originated in Changshan, Zhejiang Province as a natural cross between pummelo and sweet orange, is mainly grown in Zhejiang Province.

Main Characters: Tree vigorous, spheroid, branches somewhat erect. Fruit pyriform to oblate, orange-yellow, 9.6cm×4.2cm in size, apex with prominent or faint areole ring, rind rough or smooth, easily peeling. TSS 11.0%, acid 1.0%~1.3%. 10~40 seeds per fruit, 3~4 seeds in less-seedy variety, occasionally seedless. Pulp is moderate sweet and sour with mild bitterness, fresh and rich taste. Fruit has relatively fine quality, matures in middle to late November and stores well. The variety is quite productive and comparatively cold-resistant. Common rootstock is Trifoliate orange.

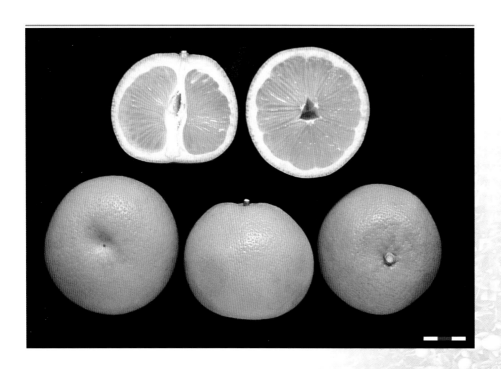

4. 鸡尾葡萄柚

来源与分布：鸡尾葡萄柚是暹罗甜柚和弗鲁亚橘的一个杂交种，20世纪50年代美国加利福尼亚大学河滨分校杂交培育。20世纪90年代华中农业大学从美国引进。目前，在湖北、江西、浙江、广东和云南等地均表现丰产，容易栽培。在甜橙适宜区品质较好。

主要性状：树势强健，树冠圆头形，树姿开张；幼树树冠扩张快、稍直立，进入结果期后渐开张，枝条长势强。叶片大小介于橙与柚之间，阔椭圆形，叶翼较小，倒披针形或线形，叶肥厚平滑，浓绿色有光泽。有叶花序和无叶花序均有，花单生或簇生，花蕾较大，白色，长椭圆形。果实大小介于橙与葡萄柚之间，果实扁圆形或圆球形，果形指数0.89，单果重380g左右，果面光滑，果皮橙黄色，果皮薄，皮厚0.42cm，海绵层白色，皮层较紧，较易剥离。果肉橙黄色，汁液多，风味爽口，略酸，囊瓣13~14瓣，中心联合，不易分瓣，汁胞橙黄色，柔软多汁，甜酸适中。TSS12.0%~14.5%，酸0.7%~1.0%，每100ml果汁维生素C含量39.8mg；果实可食率72.0%~79.0%。单一栽培一般7~8粒种子／果，种子多胚；混栽种子数会增加。成熟期11月底，冬季无冻害橘区，可以留树挂果，留树过长果实会出现浮皮。初果期由于挂果较少，单果较大，果皮厚，果肉略带苦味，可溶性固形物含量低，风味淡；盛果期丰产稳产性好，单果重减小，果皮薄，风味更浓。果实耐贮性好，不易失水。以枳作砧木。

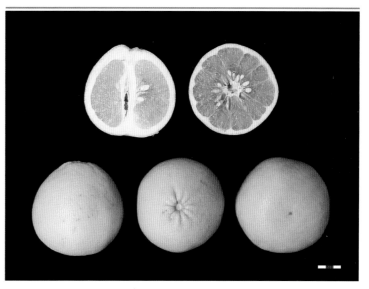

4. Cocktail Grapefruit

Origin and Distribution: Cocktail Grapefruit is a hybrid of Siamese sweet pomelo and Frua Mandarin, crossed by the University of California in Riverside in 1950s, and introduced to Wuhan by Huazhong Agricultural University in 1990s. This variety has wide adaptation, productive and easy to cultivate in Hubei, Jiangxi, Zhejiang, Guangdong and Yunnan etc. Better quality can be achieved when grown in the sweet orange production area.

Main Characters: Tree vigorous, crown spheroid and spreading. Crown spreading fast and upright in youth, open gradually after fruiting, branch strong. Leaf size between pummelo and sweet orange, broad oval, wings narrow, oblanceolate or linear. Leaf blade thick, smooth, dark-green and glossy; Leafy or leafless inflorescences, solitary or clustered; flower bud large, white, oblong. Fruit size between orange and pomelo, oblate to globose, shape index about 0.89. Fruit weight around 380g. Surface smooth, rind orange-yellow, as thin as 0.42cm. Albedo white, cortex tight, moderately easy to peel; Flesh orange-yellow in color, juicy, taste fresh, slightly sour; Segments 13~14, center connate, difficult to separate. Pulp orange-yellow, soft, juicy, moderate sweet-sour, TSS 12.0%~14.5%, TA 0.7%~1.0%, Vitamin C content 39.8mg / 100ml, edible portion 72.0%~79.0%. Seeds 7~8 when planted as singe variety, and more seeds in the mixed cultivation. Maturity at the end of November, fruits can hang on the tree in the area without frost in the winter, but develop puffy peel when hanging prolonged. Fruits are few, large, thick-peel, with slightly bitter flesh, low TSS and light flavor at the early fruit period; Fruits turn smaller, thin-peel, and flavor stronger at the full fruit period. Store well without dehydration. Trifoliate orange can be used as the rootstock.

5. 星路比葡萄柚

来源与分布：原产美国，1978年引入我国。广东、重庆等地有极少量栽培。

主要性状：树体较高大，枝细密。春梢叶片大小为7.0～8.0cm×3.2～3.5cm，长卵圆形，翼叶中等大。花中大，花粉败育。果实扁圆形，黄色至淡红色，大小为7.3～7.8cm×6.2～6.5cm，果顶平，果肉紫红色，皮肉不易分离。TSS 11.0%～12.0%，酸1.0%～1.3%，无核。果肉细嫩、汁多、味浓、微苦，品质优。成熟期11月中旬。该品种适合鲜食和加工果汁。多以枳橙或枳作砧木。

5. Star Ruby

Origin and Distribution: Star Ruby, originated in America, was introduced to China in 1978. It is rarely cultivated in Guangdong and Chongqing.

Main Characters: Tree relatively tall, branches slender and dense. Leaf 7.0~8.0cm×3.2~3.5cm in size, long ovate, petiole wing medium-large. Flower medium-large, pollen sterile. Fruit oblate, yellow to light red, 7.3~7.8cm×6.2~6.5cm in size, apex flatten, difficult to peel, seedless. TSS 11.0%~12.0%, acid 1.0%~1.3%. Pulp purple-red, tender and juicy with rich flavor and slightly bitter. Fruit matures in mid-November with fine quality. The variety is good for fresh and juice consumption. Common rootstock is Citrange or Trifoliate orange.

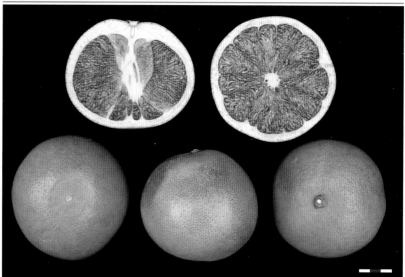

五、枸橼类
Citrons

1. 佛手

来源与分布：佛手是香橼变种，地栽以四川、广东、云南、福建、广西等地较多，盆栽主要分布于浙江金华、广东肇庆、河南鄢陵等地。

主要性状：树体矮小，树冠呈不规则圆头形，枝条披垂，有短刺。春梢叶片大小为10.0～10.7cm×4.5～5.0cm，阔椭圆形，基部楔形，先端钝尖，叶缘有锯齿，无翼叶。花中大，完全花。果实指状或拳头状长椭圆形，橙黄色，大小为9.5～10.0cm×18.0～20.0cm，果顶部分裂呈指状，果肉革质，味淡微苦，有芳香味。多作药材或观赏。

1. Fingered Citron

Origin and Distribution: Fingered citron, a variation of citron, is grown in relatively large scale in Sichuan, Guangdong, Yunnan, Fujian and Guangxi and potted mainly in Jinhua, Zhejiang Province, Zhaoqing, Guangdong Province and Yanling, Henan Province.

Main Characters: Tree dwarfed, irregular spheroid; branches drooping, with short spines. Leaf 10.0~10.7cm×4.5~5.0cm in size, broad elliptical, base cuniform, apex acumiante, margin dentate, wingless. Complete flower, medium-large. Fruit fingerlike or fist like long-ellipsoid, orange-yellow, 9.5~10.0cm×18.0~20.0cm in size, apex split resembling a human hand. Pulp is leathery, slightly flavor and bitter with fragrance. It is often used as medicinal material or ornamental plant.

2. 枸橼

来源与分布：枸橼又称香橼，原产我国西南和印度。云南、四川、广东、广西、福建、湖南、湖北、浙江、西藏有零星分布。

主要性状：树体不大，树冠扁圆形，枝节间疏长且壮，嫩枝紫红色。春梢叶片大小为9.0～12.5cm×5.0～7.0cm，阔椭圆形，叶缘波状，锯齿明显，叶柄与叶片间不具隔痕，无翼叶。花大，有花序，花瓣常带紫红色。果实椭圆形，橙黄色，大小为7.7～8.0cm×9.5～10.0cm。果顶渐尖有乳头凸起，果蒂部有5～6条沟纹，果皮不易剥离，果肉白色，汁少，酸苦味，种子大而多，20～25粒/果。多作药材或观赏。

2. Citron

Origin and Distribution: Citron, also named as Xiangyuan, originated in southwestern China and India and is scattered in Yunnan, Sichuan, Guangdong, Guangxi, Fujian, Hunan, Hubei, Zhejiang and Tibet.

Main Characters: Tree medium, crown oblate, branches thick with long internode distance, young twigs purple-red. Leaf 9.0~12.5cm×5.0~7.0cm in size, broad elliptical, margin sinuate with prominent dentate, wingless, no visible articulation between leaf blade and petiole. Flower large, inflorescence, petal usually purple-tinged. Fruit elliptsoid, orange-yellow, 7.7~8.0cm×9.5~10.0cm in size; apex blunt-pointed with pronounced mammilla; base striped with 5~6 furrows; difficult to peel. Pulp is white, less juicy and tastes sour-bitter. Seeds is large-sized and numerous, 20~25 seeds per fruit. Citron is often used as medicinal herb or ornamental plant.

六、柠檬/檬檬
Lemons/Limonias

1. 白檬檬

来源与分布：白檬檬又名白柠檬、土柠檬等，原产我国，可能是柠檬与宽皮柑橘的自然杂交种。在广西、广东、台湾、云南、贵州等地有野生分布。

主要性状：树体矮小，树冠半圆形，枝条细长披散，有短刺。春梢叶片大小为6.8～7.2cm×2.8～3.0cm，叶缘锯齿浅。花中大。果实圆球形，橙黄色，大小为3.0～3.4cm×2.8～3.2cm。TSS 10.0%～11.0%，酸4.8%～5.0%，种子5～15粒/果，成熟期11～12月。该品种的适应性与红皮酸橘同，广西用作宽皮柑橘的砧木。

1. White Limonia

Origin and Distribution: White limonia, also named as Bainingmeng, Tuningmeng etc., originates in China and perhaps is a natural cross of Lemon and Mandarin. There are distributions in Guangxi, Guangdong, Taiwan, Yunnan, and Guizhou.

Main Characters: Tree short, semi-spheroid, branches slender, long and drooping, with short spines. Leaf 6.8~7.2cm×2.8~3.0cm in size, margin shallowly dentate. Flower medium-large. Fruit spheroid, orange-yellow, 3.0~3.4cm×2.8~3.2cm in size. TSS 10.0%~11.0%, acid 4.8%~5.0%, 5~15 seeds per fruit. Maturity period ranges from November to December. The adaptability of the variety is similar to the Hongpi sour mandarin. It is commonly used as rootstocks of Loose-skin mandarins in Guangxi.

2. 北京柠檬

来源与分布： 北京柠檬又名香柠檬，可能是柠檬与橘的杂交种。原产我国。浙江、重庆、广东有少量栽培。

主要性状： 树冠圆头形，开张，枝条细长，有短刺。春梢叶片大小为7.2～8.5cm×4.0～4.5cm，椭圆形，先端渐尖，基部楔形，叶缘有锯齿，叶柄很短。花大，带紫色，一年开花多次。果实椭圆形，橙色，果实比尤力克柠檬稍大，果顶乳突短而略小，果皮光滑。TSS 6.0%，酸4.1%，种子4～5粒/果，味酸。成熟期11月上旬。果皮薄，果实耐贮性稍差。该品种抗寒力较强，易感流胶病。

2. Meyer Lemon

Origin and Distribution: Meyer lemon, also named as Xiangningmeng, perhaps originated from the cross of lemon and mandarin in China. There are some plantings of Meyer lemon in Zhejiang, Chongqing and Guangdong.

Main Characters: Tree spheroid, spreading; branches slender and long, with short spines. Leaf 7.2~8.5cm×4.0~4.5cm in size, elliptical, apex acuminate, base cuniform, margin dentate, petiole very short. Flower large, purple tinged, everflowering. Fruit ellipsoid, orange, somewhat larger than Eureka; apex with short and somewhat small mammilla; rind thin and smooth. TSS 6.0%, acid 4.1%; 4~5 seeds per fruit. Fruit matures in early November and stores somewhat poorly. The variety is relatively resistant to cold, susceptible to Gummosis Disease.

3. 粗柠檬

来源与分布：原产印度，是枸橼与柠檬的杂交种。重庆、湖北、广东有引种。

主要性状：树势强，树姿开张，树冠不规则圆头形。春梢叶片大小为7.7～9.1cm×3.1～4.4cm，椭圆形。单花或总状花序，花蕾紫红色。果实椭圆形，橙黄色，大小为5.3～5.7cm×5.0～5.9cm，果顶部有明显乳头凸起，果蒂部有数条放射沟纹，果面粗糙，含酸3.5%。种子25～30粒/果。作砧木用。

3. Rough Lemon

Origin and Distribution: Rough lemon, originated in India, is a cross of citron and lemon. It was introduced into Chongqing, Hubei and Sichuan.

Main Characters: Tree vigorous, spreading, irregular spheroid. Leaf 7.7~9.1cm×3.1~4.4cm, elliptical. Solitary flower or raceme, flower bud purlpe-red. Fruit ellipsoid, orange-yellow, 5.3~5.7cm×5.0~5.9cm in size, apex with pronounced nipple, base striped with several radial furrows, surface rough, acid 3.5%. 25~30 seeds per fruit. It is commonly used as rootstock.

4. 红檬檬

来源与分布：红檬檬又名广东柠檬，原产我国，可能是柠檬与宽皮柑橘的自然杂交种。在广东、湖南、台湾、贵州、云南、四川、重庆等地有分布。

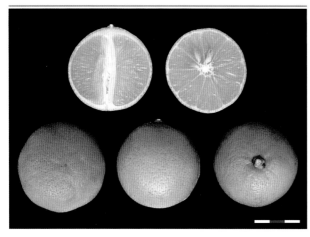

主要性状：树体矮小，树冠半圆形，枝条细长披垂，嫩梢叶片浅紫色。春梢叶片大小为7.2～8.0cm×3.2～3.6cm，卵圆形或长椭圆形，叶缘锯齿较明显，翼叶线状。花中大。嫩梢和花蕾紫红色。果实近圆形，橙红色或朱红色，光滑，大小为3.8～4.0cm×3.6～3.8cm，果顶乳凸不甚明显，皮薄，不易剥离。TSS 10.0%～11.0%，酸5.0%～5.5%，种子16～17粒/果，成熟期11～12月。该品种抗衰退病，耐盐性和耐湿性强，不抗裂皮病，根浅不耐旱易衰老，是广东、广西稻田栽培柑橘的砧木。

4. Red Limonia

Origin and Distribution: Red limonia, also called as Guangdong lemon, originated in China and perhaps is a natural cross of Lemon and Mandarin. It is distributed in Guangdong, Hunan, Taiwan, Guizhou, Yunnan, Sichuan and Chongqing.

Main Characters: Tree short, semi-spheroid, branches slender, long and drooping, young leaves and twigs light-purple. Leaf 7.2~8.0cm×3.2~3.6cm in size, ovate to long-elliptical, margin dentate conspicuous, petiole wing linear. Flower medium-large. Flower buds tinged with purple-red. Fruit nearly spheroid, orange-red to deep red, surface smooth, 3.8~4.0cm×3.6~3.8cm in size, apex nipple inconspicuous, pericarp thin, difficult to peel. TSS 10.0%~11.0%, acid 5.0%~5.5%. 16~17 seeds per fruit. Maturity period ranges from November to December. The variety is resistant to Citrus tristeza virus, and tolerant to saline soil and humid condition, but susceptible to Citrus exocortis viroid. It is sensitive to drought stress due to shallowly distributed roots, which causes tree decline. It is commonly used as rootstocks of paddyfield citrus in Guangdong and Guangxi.

5. 尤力克柠檬

来源与分布：原产美国。我国20世纪30年代引入，现四川栽培最多，广东、云南也有发展，台湾、福建、广西等地有少量栽培。

主要性状：树势强健，树姿开张，枝细长而稀疏，刺少而短。春梢叶片大小为11.5～14.0cm×4.5～5.5cm，椭圆形，叶缘波状，有浅锯齿，翼叶线状。嫩梢、花蕾均带紫色。花大，一年四季开花。果实椭圆形，淡黄色，单果重120～180g，果顶部有乳头凸起，蒂部钝圆，有放射沟纹，皮厚而粗。果汁多，香气浓。TSS 7.5%～8.5%，酸6.0%～7.5%。该品种宜制果汁和提炼精油，品质上等。

5. Eureka Lemon

Origin and Distribution: Eureka lemon originated in America and was introduced to China in 1930s. Currently, it is cultivated mostly in Sichuan, moderately in Guangdong and Yunnan, and limitedly in Taiwan, Fujian and Guangxi.

Main Characters: Tree vigorous, spreading, branches slender, long and sparse, spines less and short. Leaf 11.5~14.0cm×4.5~5.5cm in size, elliptical, margin sinuate, shallowly dentate, petiole wing linear. Young twigs and buds purple-tinged. Flower large, everflowering. Fruit ellipsoid, light yellow, weight 120~180g. Apex mammiform. Base obtuse to rounded, striped with radial grooves and ridges. Pericarp thick and rough. Pulp juicy, very fragrant; TSS 7.5%~8.5%, acid 6.0%~7.5%. Its quality is superior for juice and essential oil processing.

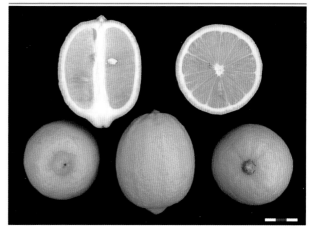

6. 云柠1号柠檬

来源与分布：云南省农业科学院热带亚热带经济作物研究所从尤力克柠檬芽变中选育。2011年通过云南省农作物品种审定委员会认定。

主要性状：树形紧凑，树冠圆头形，枝梢密度适中，生长势适中。单叶，嫩叶紫红色，叶卵圆形，叶尖短尖，叶基为广楔形，叶宽5.8cm，叶长11.9cm，无翼叶，叶缘为圆齿状。花腋生，丛生状，两性花，花粉数量多，花瓣紫红色，花丝部分联合，花柱直立，一年多次开花。单果重145g，椭圆形，有乳突，无果顶印圈，花柱脱落，果基呈颈领状。果皮黄色，厚0.83cm，难剥离，果面光滑，油胞凹陷。果心充实，内果皮白色，果肉黄白色，囊瓣约10瓣，种子10～45粒/果，汁多肉脆，香气好，TSS 8.5%、总酸7.1%、每100ml果汁维生素C含量49mg。

在云南，一般以枳为砧木，一年可抽生6～11次梢，每次间隔20～35d。1月中下旬开始出现春花蕾，2月中旬初花，3月中旬盛花，3月下旬终花，2～3月出现2次生理落花高峰，3～4月出现生理落果。通过套袋技术45d左右可转色，8月中旬可上市，自然条件下11月下旬转色。定植当年即可挂果，第三年进入丰产稳产期，产量达到60t/hm^2，优质果率达70%。

适宜于云南柑橘产区，海拔650～1 500m，年均温17～22℃，年降水量750～1 600mm，>10℃活动积温5 200～7 300℃，pH5.5～6.8，土层厚度>100cm，地下水位<100cm的沙壤土、壤土上种植。

6. Yunning No.1 Lemon

Origin and Distribution: Yunning No.1 Lemon was a bud sport selected from Eureka Lemon by the Institute of Tropical and Subtropical Economic Crops, Yunnan Academy of Agricultural Sciences and registered in Yunnan Province in 2011.

Main Characters: Tree compact, crown spheroid, moderate shoot density, medium growth potential; simple leaves, young leaves purple-red; oval leaves with short pointed tips and broad wedge-shaped bases, 5.8cm wide and 11.9cm long, no winged petiole, dentate leaf margin. It blooms multiple times a year and flowers are axillary, fascicular and bisexual, with much fertile pollen, purple-red petals, partially joint filaments and upright styles. The fruit weight 145g on average, elliptic, apex mammiform without areole ring, style detached and fruit base collar-like. The pericarp is yellow, 0.83 cm thick and difficult to peel. Fruit surface is smooth with depressed oil glands. The fruit has a solid core, white endocarp and yellowish-white pulp. The fruit has around 10 segments, 10~45 seeds, pulp juicy and crisp, good aroma, TSS 8.5%, TA 7.1%, Vitamin C 49mg/100ml.

In Yunnan, Trifoliate orange is generally used as the rootstock. Branching happens 6 to 11 times a year, with an interval of 20 to 35 days. Spring bud appears in mid to late January, flowering starts in mid-February, full blossom in mid-March and ends in late March. The peak of physiological flower drop appear twice in February to March, and the physiological fruit drop happens in March to April. Fruits change color 45 days after bagging and can go on the market in mid-August, while color changes in the late November for naturally growing fruits. In Yunnan, the newly planted tree can set fruit in the first year, and reaches a normal production of 60t/hm^2 in the third year with 70% high-quality fruit rate.

It is suitable for citrus production area in Yunnan Province with an altitude of 650~1 500m, an annual average temperature of 17~22℃, annual rainfall of 750~1 600mm, and the accumulated temperature of 5 200~7 300 above 10℃; on sandy loam or loam soil with a pH value of 5.5 to 6.8, soil layer thickness above 100cm, and groundwater level below 100cm.

七、枳以及枳和柑橘属的杂种

Trifoliate Oranges and Its Hybrids with Citrus Genus

1. 枳

来源与分布： 枳又名枸橘、枳壳、雀不站和铁篱笆。原产我国，湖北、河南、山东、安徽、福建、江西、江苏等省有分布。

主要性状： 落叶性灌木或小乔木。树姿开张，枝条硬刺多且长。叶为三出复叶。一年多次开花，花小，完全花。果实圆球形或扁圆形，暗橙黄色，大小为3.8～4.0cm×3.6～4.2cm，果顶微突，有印圈，果面密被茸毛，较粗糙，皮包着紧，难剥离。味酸苦涩不堪食用。种子20～35粒/果，9～10月成熟。该品种在国内有大叶大花枳、小叶小花枳等类型，以大叶大花型比较优良。枳冬季落叶，有一定的休眠期。是我国多数柑橘产区的柑橘砧木，但南亚热带地区不宜作蕉柑、粤英甜橘、新会甜橙等的砧木。

1. Trifoliate Orange

Origin and Distribution: Trifoliate orange, also called as Gouju, Zhike, Quebuzhan and Tieliba, is originated in China and distributed in Hubei, Henan, Shandong, Anhui, Fujian, Jiangxi, Jiangsu and etc.

Main Characters: Deciduous shrub or small arbor. Tree spreading, branches long with hard and long spines. Compound leaf, trifoliate. It can be flowring several times in one year. Complete flower, small. Fruit spheroid to oblate, dark orange-yellow, 3.8~4.0cm×3.6~4.2cm in size, apex slightly convex with areole ring. Rind surface densely pubescent, relatively rough, tightly adherent and difficult to peel. Tastes sour-bitter and inedible. 20~35 seeds per fruit. Maturity period ranges from September to October. The Trifoliate orange contains large-leaf-large-fruit, small-leaf-small-fruit and other lines, among which the large-leaf-large-fruit is a relatively better rootstock. Trifoliate orange defoliates in winter and has a definite period of dormancy. It is used as rootstock of citrus in most citrus producing areas in China, but not a suitable rootstock for Jiaogan, Yueyingtianju, Xinhui sweet orange or other citrus growing in southern subtropical regions.

2. 枳橙

来源与分布：枳与橙类的属间杂种，我国秦岭以南、淮河以南以及长江流域各省均有分布。

主要性状：树势强，较直立，枝多刺。以三小叶组成的复叶为主，也有两小叶组成的复叶和单复叶。果实圆球形，大小为4.8cm×4.7cm，果皮橙黄色，较粗糙或光滑，果肉酸略带苦麻味。种子20～30粒/果。该品种根系发达，开花期比枳晚，比橙类稍早，耐寒力仅次于枳，抗病力强，半矮化，是多数柑橘品种早结丰产的良好砧木。我国有南京枳橙、黄岩枳橙、永顺枳橙、有毛枳橙等，从国外引进的有卡里佐枳橙、特洛亚枳橙，均可作宽皮柑橘和甜橙的砧木。目前生产上使用较多的是美国培育的卡里佐枳橙。

卡里佐枳橙　Carrizo

2. Citrange

Origin and Distribution: It is a generic hybrid between *Poncirus trifoliata* and *Citrus sinensis*. Citranges are distributed in the south of the Qinling Mountain, the south of the Huaihe River and the Yangtze River regions.

Main Characters: Tree vigorous, relatively erect, branches thorny. Compound leaf, mostly trifoliate, occasionally bifolioate or unifoliolate. Fruit spheroid, 4.8cm×4.7cm in size, pericarp orange-yellow, relatively rough or smooth, pulp taste sour with slight bitterness. 20~30 seeds per fruit. The root system is well-developed. Anthesis is later than common trifoliate orange and slightly earlier than orange. Less cold-hardy than trifoliate orange, highly resistant to pathogens, semi-dwarfing. It is a good rootstock on which most citrus varieties become precocious and productive. There are Nanjingzhicheng, Huangyanzhicheng, Yongshunzhicheng, Maozhicheng and some other citrangs in China. Introduced citranges from abroad are Carrizo and Troyer. All these citranges can be used as rootstocks of Loose-skin mandarins and Sweet oranges. The widely used citrange in production is Carrizo, a variety developed in USA.

3. 枳柚

来源与分布：葡萄柚和枳的属间杂种。广西、湖北等地有分布。

主要性状：与枳橙相同，叶片有3小叶、2小叶及单身复叶三种，幼枝叶具稀疏小短茸毛，我国枳柚杂种不多，主要有广西1号、广西4号以及从国外引进的施文格枳柚。该品种作为宽皮柑橘和哈姆林甜橙砧木表现良好。与默科特及柠檬等嫁接表现不亲和。主要缺点是接穗过旺地生长，常导致橙和宽皮柑橘早衰，以及冬季出现叶片黄化等症状。

3. Citrumelo

Origin and Distribution: It is a generic hybrid of Trifoliate orange and Grapefruit, and is distributed in Guangxi and Hubei, etc.

Main Characters: Leaves similar to Zhicheng with three types: trifoliate, bifoliate and unifoliate. Young branches and leaves sparsely puberulent. The numbers of Citrumelos in China is less, mainly including Guangxi No.1, No.4 and the introduced Swingle citrumelo. Citrumelo exhibits good performance when used as rootstock of mandarins and Hamlin sweet orange, but is incompatible with Murcutt and Lemon. The main disadvantage using Citrumelo as rootstock is that the scion varieties on it grow too fast, causing early tree decline of scion cultivars and leaf yellowing in winter.

施文格枳柚　Swingle

八、金柑
Kumquats

1. 长寿金柑

来源与分布：长寿金柑又名寿星橘、月月橘、公孙橘，可能是金柑与橘的杂交种。各柑橘产区有零星栽培。

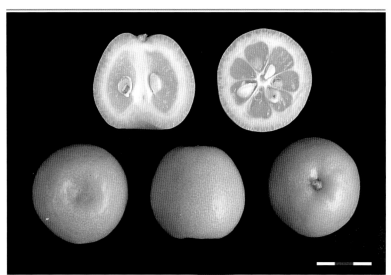

主要性状：树冠圆头形，枝条节间短，无刺。春梢叶片大小为5.2～5.6cm×3.1～3.3cm，短椭圆形。花小，完全花。果实倒卵形，橙黄色，有光泽，大小为3.4～3.5cm×3.3～3.4cm。TSS 14.0%～15.0%，酸1.0%～1.5%，种子4～5粒/果。该品种丰产性好，果皮脆甜，果肉酸，鲜食不佳，可加工制蜜饯。是广东盆栽观赏柑橘的搭配品种。

1. Changshou Kumquat

Origin and Distribution: Changshou kumquat (*F. obovata*), also named as Shouxingju, Yueyueju, Gongsunju, is probably a cross of Jingan and Mandarin. There are small-scale cultivations in all principal citrus producing areas.

Main Characters: Tree spheroid topped, branches with short internode distance and spineless. Leaf 5.2~5.6cm×3.1~3.3cm in size, short-elliptic. Complete flower, small. Fruit obovoid, orange-yellow, glossy, 3.4~3.5cm×3.3~3.4cm in size. TSS 14.0%~15.0%, acid content 1.0%~1.5%. 4~5 seeds per fruit, rind crisp and sweet, pulp sour, eating quality not so high. Good for making candy kumquat. The variety is very productive and is one of potted ornamental citrus in Guangdong Province.

2. 脆蜜金柑

来源与分布：脆蜜金柑为滑皮金柑的优良变异，2007年在广西融安县大将镇滑皮金橘园发现，2014年通过广西农作物品种审定委员会审定。

主要性状：树势强，树冠较开张，萌芽力中等，成枝力强。结果母枝以春梢为主。自然坐果率低，保花保果后丰产。果实椭圆形至圆球形，果面光滑、金黄色至橙红色，单果重20.5g，大小为2.8～4.6cm×3.1～4.9cm，果形指数0.97～1.22，少核或无核，TSS19.1%～25.0%，总糖15.8%～20.5%，酸0.1%～0.2%，每100ml果汁维生素C含量21.6～35.3mg，固酸比126.3～161.1，风味浓甜，化渣，品质优。在广西成熟期11月下旬至12月中旬。耐旱，耐寒，耐贮藏，抗柑橘溃疡病。12月上旬树冠覆膜可避雨防裂果落果，延迟采收。适宜在金柑种植区推广。

2. Cuimi Kumquat

Origin and Distribution: Cuimi kumquat is a variation of Huapi kumquat, found in Dajiang, Rong' an County, Guangxi in 2007 and registered in 2014.

Main Characters: Trees vigorous, crown somewhat open, budding capacity medium, branching capability strong. Fruit twigs are mainly spring shoots. Natural fruit setting rate low, yield high when protecting flowering and fruits from dropping. Fruit elliptic or spherical, peel smooth, golden yellow to orange-red. Average fruit weight 20.5g, size in 2.8~4.6cm×3.1~4.9cm, fruit shape index 0.94~1.22, few seeds or seedless, TSS 19.1%~25.0%, total sugar 15.8%~20.5%, TA 0.1%~0.2%, Vitamin C 21.6~35.3mg/100ml juice. Flavor rich in sweet, melting, quality superior. Maturity in late November to mid-December in Guangxi. Cold hardy, drought-tolerant, resistant to citrus canker. Fruits are storage durable. In early December, covering film on the canopy can protect the fruits from dropping and cracking and delayed harvest induced by rain. Suitable for kumquat cultivation area.

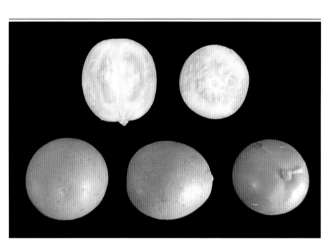

3. 桂金柑1号

来源与分布：桂金柑1号是由阳朔金柑实生系选育出的早熟金柑新品种，2016年通过广西农作物品种审定委员会审定。

主要性状：成枝力中等。1年开花3~4次，花单生、双生或簇生。果实椭圆形，果皮橘黄色，有光泽，油胞明显，果肉淡黄色，味清甜有辛辣味，单果重24.4~32.5g，果形指数1.14~1.24，种子3.4~6.4粒/果，可食率97.1%~98.1%。每100ml果汁维生素C含量28.3~53.0mg，酸0.2%~0.5%，全糖10.3%~16.6%，TSS12.6%~18.6%，品质上等。果实生育期160~180d，在桂林阳朔11月上中旬至12月中旬成熟；耐寒、耐旱、耐瘠薄土壤，适应性广，抗病性强。定植后第三年开花结果，丰产稳定性好。适宜在广西桂林、柳州等地以及生态条件相似地区栽培。

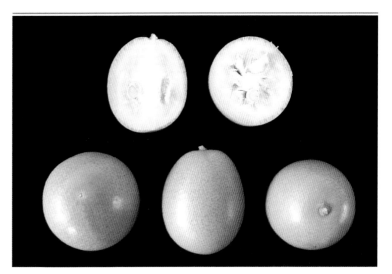

3. Guijingan No.1 Kumquat

Origin and Distribution: Guijingan No.1 kumquat is an early-season cultivar selected from seedlings of Yangshuo kumquat, registered in Guangxi in 2016.

Main Characters: Fruit ellipsoid, rind orange-yellow, glossy, oil glands immersed. Pulp light yellow, taste sweet, spicy, peppery, fruit weight 24.4~32.5g, shape index 1.14~1.24, seed 3.4~6.4 per fruit, edible portion 97.1%~98.1%. TSS 12.6%~18.6%, TA 0.2%~0.5%, total sugar 10.3%~16.6%, Vitamin C 28.3~53.0mg/100ml, quality superior. Fruit development period 160~180 days, maturity in early November to mid-December, in Yangshuo, Guilin, Guangxi. Flowering 3 to 4 times per year, singly, twin or cluster. Branching capacity medium. Cold hardy, drought-tolerant, barren-resistant, widely adaptable, disease resistant. Stably prolific, this variety can bear fruits from the third year after plantation. Suitable for Guilin, Liuzhou and other places in Guangxi with similar ecological conditions.

4. 桂金柑2号

来源与分布：桂金柑2号是由阳朔金柑实生系选育出的晚熟金柑新品种，2016年通过广西农作物品种审定委员会审定。

主要性状：果实椭圆形，橙红色，光滑，油胞平生，果顶圆钝；单果重24.3～34.1g，大小为3.4～3.7cm×3.9～4.4cm，果形指数1.11～1.20，可食率97.8%～99.0%，每100ml果汁维生素C含量32.0～54.4mg，酸0.2%～0.5%，TSS15.2%～19.9%，全糖12.8%～15.2%。种子2.1～5.5粒/果，多胚，子叶淡绿色。风味浓郁，有微淡刺鼻味，果实汁多、较化渣，品质优。果实生育期180～195d，在桂林阳朔12月中下旬至翌年1月中旬成熟。定植后第三年开花结果，丰产性好。适宜在广西桂林、柳州等地以及生态条件相似地区栽培。

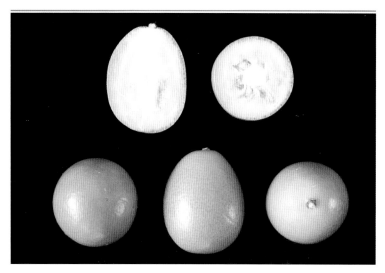

4. Guijingan No.2 Kumquat

Origin and Distribution: Guijingan No.2 kumquat is a late-season cultivar selected from seedlings of Yangshuo kumquat, registered in Guangxi in 2016.

Main Characters: Fruit ellipsoid, rind orange-red, glossy, smooth, oil glands inconspicuous, apex round and blunt. Fruit weight 24.3~34.1g, size in 3.4~3.7cm×3.9~4.4cm, fruit shape index 1.11~1.20. Edible portion 97.8%~99.0%, total sugar 12.8%~15.2%, TSS 15.2%~19.9%, TA 0.2%~0.5%. Vitamin C 32.0~54.4mg/100ml. 2.1~5.5 seeds per fruit, polyembryonic, cotyledons light green. Flavor rich, slightly pungent, juicy, melting, quality superior. Fruit development period 180~195 days. Maturity from mid-December to the following mid-January in Yangshuo, Guilin. Stably prolific, this variety can bear fruits from the third year after plantation. Suitable for Guilin, Liuzhou and other places in Guangxi with similar ecological conditions.

5. 金弹

来源与分布：金弹又名金柑、长安金橘、融安金橘、龙溪金柑、遂川金柑、上坪金柑等。可能是罗浮和罗纹的杂交种。广西、福建、江西、浙江、广东栽培较多。

主要性状：树冠圆头形或倒椭圆形，枝条细而密生，有短刺。春梢叶片大小为5.5～7.3cm×2.3～3.2cm，呈椭圆形，叶正面深绿色，背面灰白色。花小，完全花。果实椭圆形或卵状椭圆形，橙黄色或金黄色，有光泽，大小为2.7～3.0cm×3.0～3.4cm。TSS 15.0%～17.0%，酸0.4%～0.5%。果皮甘甜，肉质味甜，适合鲜食。种子3～5粒/果。是市面销售的鲜食金柑品种，也有加工成糖水罐头及蜜饯。该品种适应性强，丰产稳产，特耐溃疡病。

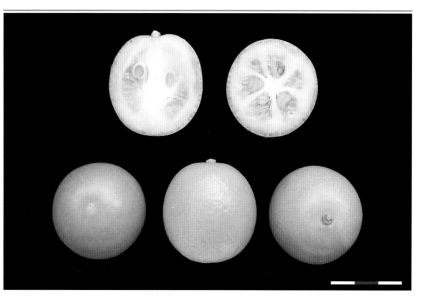

5. Jindan

Origin and Distribution: Jindan (*F. crassifolia*), also named as Jingan, large round kumquat, Changanjinju, Ronganjinju, Longxijingan, Suichuanjingan, and Shangpingjingan, is probably a hybrid between Luofu and Luowen and mainly grows in Guangxi, Fujian, Jiangxi, Zhejiang and Guangdong.

Main Characters: Tree spheroid or obovoid, branches dense and slender with short spines. Leaf 5.5~7.3cm×2.3~3.2cm in size, elliptic, upper surface dark green, lower surface pale gray. Complete flower, small. Fruit ellipsoid to ovoid, orange-yellow or golden yellow, surface glossy, 2.7~3.0cm×3.0~3.4cm in size. TSS 15.0%~17.0%, acid content 0.4%~0.5%. Fruit tastes very sweet and is excellent for fresh consumption. 3~5 seeds per fruit. It is the variety commonly found in fresh market and also processed as canned kumquat and candied kumquat. It presents a strong adaptability with consistent and productive yield and highly resistant to citrus canker.

6. 金豆

来源与分布：金豆又名山金柑、山金豆、山橘、香港金柑等，广东、广西、福建、浙江、湖南、江西、香港等山地野生。本种是柑橘类中唯一的天然四倍体，染色体数为36。

主要性状：矮小灌木，树势弱，树冠较直立，枝纤细多刺。春梢叶片大小为4.0～4.3cm×1.5～2.0cm，呈卵状椭圆形。花小，完全花。果实圆形，金黄色或橙红色，大小为1.0～1.3cm×0.9～1.2cm，种子3～4粒/果。该品种生长慢，耐寒耐瘠，果肉及果汁极少，味酸苦，尚未进行经济开发和利用。

6. Jindou

Origin and Distribution: Jindou (*F. hindsii*), also named as Shanjingan, Shanjindou, Shanju or Hongkong jingan, grows wildly in mountains of Guangdong, Guangxi, Fujian, Zhejiang, Hunan and Hongkong. This species, with 36 chromosomes, is the only natural tetraploid among citrus.

Main Characters: Tree small shrub, low-vigorous, comparatively erect, branches slender with many spines. Leaf 4.0～4.3cm×1.5～2.0cm in size, ovate to elliptic. Complete flower, small-sized. Fruit small, spheriod, golden yellow or orange-red, 1.0～1.3cm×0.9～1.2cm in size, 3～4 seeds per fruit. Slow in growth, tolerant to cold and leanness. The proportion of pulp is less with lacking in juice and acid-bitter taste. It has not been used commercially.

7. 罗浮

来源与分布： 罗浮又名金枣、牛奶金柑、长实金柑，浙江、江西、广东、福建有分布。

主要性状： 树冠半圆形或圆头形，枝条细密，稍直立。春梢叶片大小为7.0～7.5cm×2.2～2.7cm，呈长椭圆形或阔披针形，先端尖，叶面深绿色，叶背灰白色，边缘扭曲有波状。花小，完全花。果实长椭圆形或长卵形，橙黄色，大小为2.0～2.2cm×2.9～3.2cm。TSS 12.0%～13.0%，酸0.5%～0.6%，种子4～5粒/果。果皮甘甜，果肉较酸，可鲜食，品质不及金弹，可加工制蜜饯。广东盆栽观赏柑橘的搭配品种，特耐溃疡病。

7. Luofu

Origin and Distribution: Luofu (*F. margarita*), also named as Nagami, Oval kumquat, Jinzao, Niunaijingan and Changshijingan, is distributed in Zhejiang, Jiangxi, Guangdong and Fujian.

Main Characters: Tree spheroid or semi-spheroid, branches slender and dense, somewhat erect. Leaf 7.0~7.5cm×2.2~2.7cm in size, long-elliptic or broad lanceolate, apex acute, upper surface dark green, lower surface pale gray, margin twist, sinuate. Complete flower, small-sized. Fruit long-ellipsoid or long-ovoid, orange-yellow, 2.0~2.2cm×2.9~3.2cm in size. TSS 12.0%～13.0%, acid content 0.5%~0.6%, 4~5 seeds per fruit. Rind tastes sweet; pulp relatively acidic; eating quality is not as good as Jindan kumquat. Fruit can be candied. It is highly resistant to citrus canker and is an alternative variety used for potted ornamental citrus in Guangdong Province.

8. 罗纹

来源与分布：罗纹又名圆金柑、金橘，浙江栽培较多。

主要性状：灌木，枝有少刺，春梢叶片长卵形，大小为6.5～7.0cm×2.7～3.0cm。果实球形，橙黄色，大小为2.0～2.2cm×2.1～2.3cm，油胞大而凸起。TSS 10.3%，酸1.2%～1.3%。种子1～3粒/果，肉微硬，鲜食不如金弹。成熟期10月底至11月初。

8. Luowen

Origin and Distribution: Luowen (*F. japonica*), also named as Round kumquat, Yuanjingan and Jinju, mainly grows in Zhejiang.

Main Characters: Tree shrub, branches less thorny. Leaf 6.5~7.0cm×2.7~3.0cm in size, long-ovate. Fruit spheroid, orange-yellow, 2.0~2.2cm×2.1~2.3cm in size, oil glands large and conspicuous. TSS 10.3%, acid content 1.2%~1.3%, 1~3 seeds per fruit. Pulp slightly firm, eating quality is not as good as Jindan. Maturity period ranges from late October to early November.

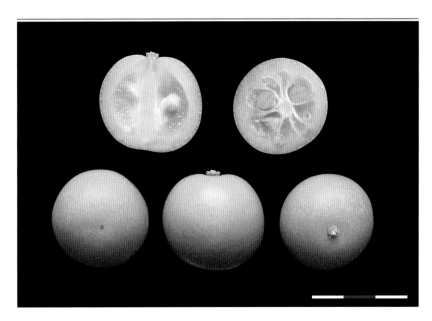

9. 四季橘

来源与分布：四季橘原产广东，可能是金柑和宽皮柑橘的杂交种。分布在广东、广西、福建等地。

主要性状：树冠圆头形，枝条细长，有短刺。春梢叶大小为5.7～6.5cm×3.2～3.6cm，呈卵圆形或卵状椭圆形，叶缘锯齿稍明显，翼叶小。花小，完全花。果实扁圆形，橙黄色，大小为2.2～2.3cm×2.0～2.1cm，果顶果蒂部微凹，果皮易剥离。TSS 10.0%～10.5%，酸2.5%～3.0%，味极酸，不堪食用。种子4～5粒/果。该品种丰产稳产，是广东盆栽观赏橘柑的主栽品种。果实可加工制蜜饯。

9. Sijiju

Origin and Distribution: Sijiju (*C. madurensis*), originated in Guangdong, is probably a hybrid between Jingan and Mandarin. It is distributed in Guangdong, Guangxi and Fujian, etc.

Main Characters: Tree spheroid, branches slender and long, with short spines. Leaf 5.7~6.5cm×3.2~3.6cm in size, ovate to elliptic, margin slightly dentate, petiole wing small. Complete flower, small-sized. Fruit oblate, orange-yellow, size in 2.2~2.3cm×2.0~2.1cm, apex and base slightly depressed, easy peeling. TSS 10.0%~10.5%, acid content 2.5%~3.0%. 4~5 seeds per fruit. Its yield is consistent and productive. It is the principal variety of ornamental citrus in Guangdong Province. Fruit can be processed as candy kumquat.

主要参考文献
Main References

陈竹生, 等, 1993. 中国柑橘良种彩色图谱[M]. 成都: 四川科学技术出版社.

华南农学院, 1979. 果树栽培学各论:南方本:下册[M]. 北京: 农业出版社.

李道高, 1996. 柑橘学[M]. 北京: 中国农业出版社.

农业部农业司, 中国农业科学院柑橘研究所, 1992. 中国名特优柑橘及其栽培[M]. 上海: 上海科学技术出版社.

彭成绩, 等, 1996. 广东柑橘图谱[M]. 广州: 广东科技出版社.

沈兆敏, 等, 1992. 中国柑橘技术大全[M]. 成都: 四川科学技术出版社.

石健泉, 1988. 广西柑橘品种图册[M]. 南宁: 广西人民出版社.

徐建国, 2003. 柑橘优良品种及无公害栽培技术[M]. 北京: 中国农业出版社.

俞德浚, 1979. 中国果树分类学[M]. 北京: 农业出版社.

张秋明, 等, 1992. 脐橙高产栽培技术[M]. 长沙：湖南科学技术出版社.

章文才, 1977. 柑橘生产技术与科学实验[M]. 北京: 科学出版社.

中华人民共和国农业部, 2007. 2006中国农业统计资料[M]. 北京: 中国农业出版社.

Reuther W, Batchelor LD, Webber HJ (eds), 1967. The Citrus Industry: Vol.I: History, World Distribution, Botany, and Varieties[M]. California, USA: University of California Press.

Saunt J, 2000. Citrus Varieties of the World: An Illustrated Guide[M]. Norwich, England: Sinclair International.

中国柑橘品种中文名称索引
Index to Citrus Varieties in China (in Chinese)

1232 橘橙	97	长叶晚橙	133	朱栾	201
HB 柚	202	长寿金柑	252	朱橘	61
		化州橙	144	先锋橙	159
二画				伏令夏橙	137
八月橘	14	**五画**		伦晚脐橙	179
九月橘	27	玉环柚	229	华柑 2 号椪柑	40
		甘平	101	华柑 4 号椪柑	42
三画		本地早	15	华柚 2 号	215
三红蜜柚	222	东 13 椪柑	37	华盛顿脐橙	177
三湖红橘	63	东江本地早	16	伊予柑	128
大三岛脐橙	168	东试早柚	207	行柑	20
大分	81	卡特夏橙	150	冰糖橙	132
大红柑	17	北京柠檬	242	安江香柚	203
大红袍	22	由良	89	兴津	93
大红甜橙	135	四季抛	226	阳山橘	59
大雪柑	164	四季橘	265	阳光 1 号橘柚	127
大雅柑	100	代代	198	红玉血橙	195
小雪柑	166	白 1 号蕉柑	72	红玉柑	107
马水橘	28	白檬檬	241	红皮酸橘	55
		处红柚	204	红肉脐橙	176
四画		市文	88	红宝石柚	213
丰彩暗柳橙	153	立间	92	红柿柑	106
天草	123	永春椪柑	47	红美人	105
韦尔金橘	57			红锦	103
云柠 1 号柠檬	247	**六画**		红韵香柑	108
五月橘	58	早玉文旦	231	红橘	21
不知火	98	早红脐橙	192	红檬檬	244
尤力克柠檬	245	早香柚	230		
少核年橘	35	早蜜椪柑	50	**七画**	
少核贡柑	69	早熟蕉柑	78	贡柑	68
日南 1 号	87	早橘	60	克里迈丁红橘	24
中柑蜜橙	167	年橘	34	园丰脐橙	190
牛肉红朱橘	62	朱红橘	64	佛手	238
长叶香橙	134	朱砂橘	66	孚优选蕉柑	73

267

沙田柚	225	金豆	261	柳橙	151
沃柑	125	金春	109	砂糖橘	53
尾张	96	金香柚	217	垫江柚	205
改良橙	139	金秋砂糖橘	113	星路比葡萄柚	237
鸡尾葡萄柚	236	金弹	260	哈姆林甜橙	143
纽荷尔脐橙	184	金煌	111	香水橙	160

八画

		乳橘	52	秋辉	122
		朋娜脐橙	185	狮头橘	200
奉节72-1脐橙	169	肥之曙	83	度尾文旦柚	208
奉晚脐橙	170	宗橙脐橙	193	宫川	91
青秋脐橙	187	试18椪柑	44	宫本	84
坪山柚	220	建柑	26		
奈维林娜脐橙	182				

九画

十画

瓯柑	79			桂金柑1号	256
软枝酸橘	56	春香	99	桂金柑2号	258
国庆1号	85	胡柚	234	桂柚1号	212
明日见	115	南3号蕉柑	74	桂夏橙1号	142
明柳甜橘	30	南丰蜜橘	31	桂橙1号	141
明柳橙	154	南柑20	95	桂橘1号	19
岩溪晚芦椪柑	46	南香	117	桃叶橙	158
罗伯逊脐橙	180	南橘	25	崀丰脐橙	178
罗纹	263	枳	248	脆蜜金柑	254
罗浮	262	枳柚	251	脐血橙	196
凯旋柑	114	枳橙	250	高橙	102
和阳2号椪柑	39	枸头橙	199	涟红	94
金乐柑	112	枸橼	240	浮廉橘	18
金兰柚	216	柳城蜜橘	32	诺瓦	118

中国柑橘品种中文名称索引
Index to Citrus Varieties in China (in Chinese)

桑麻柚	223

十一画

黄果柑	70
菠萝香柚	233
梦脐橙	181
雪柑	163
晚白柚	228
鄂柑1号椪柑	38
鄂柑2号	90
粗柠檬	243
清见	119
清家脐橙	188
梁平柚	219
隆园早	86

十二画

琯溪蜜柚	210
塔59蕉柑	75
塔罗科血橙	197
椪柑	36
椪橘	51
紫金春甜橘	67
晴姬	121
粤丰早橘	33
粤丰蕉柑	77

粤优椪柑	48
粤英甜橘	130
奥兰布兰科柚	232
奥林达夏橙	131
温州蜜柑	80
淑浦长形甜橙	162
强德勒柚	221
媛小春	129

十三画

零号雪柑	165
暗柳橙	152
锦红冰糖橙	146
锦秀冰糖橙	147
锦橙	145
新1号蕉柑	76
新生系3号椪柑	45
新会甜橙	161
满头红	61
福本脐橙	171
福罗斯特夏橙	138
福罗斯特脐橙	172
福橘	23

十四画

酸柚	227
酸橘	55
翡翠柚	209
蜜奈夏橙	156

十五画

慢橘	29
蕉柑	71
稻叶	82
德塔夏橙	136

十六画

橘红	218
橘湘元糖橙	149
橘湘珑冰糖橙	148
默科特	116
黔阳无核椪柑	43
黔阳冰糖脐橙	186
糖橙	157

二十一画

露德红夏橙	155
赣南1号脐橙	173
赣南早脐橙	175

中国柑橘品种英文名称索引
Index to Citrus Varieties in China (in English)

A

Amakusa	123
Anjiangxiangyou	203
Anliucheng	152
Asumi	115

B

Bai No.1 Jiaogan	72
Bayueju	14
Bendizao	15
Beni Madonna	105
Bingtangcheng	132
Boluoxiangyou	233

C

Cara Cara Navel Orange	176
Chandler Pummelo	221
Changshou Kumquat	253
Changye Wancheng	133
Changye Xiangcheng	134
Chuhongyou	204
Citrange	250
Citron	240
Citrumelo	251
Clementine	24
Cocktail Grapefruit	236
Cuimi Kumquat	254
Cutter Valencia Orange	150

D

Dahong Sweet Orange	135
Dahonggan	17
Dahongpao	22
Daidai	198
Daxuegan	164
Dayagan Tangor	100
Delta Valencia Orange	136

Dianjiangyou	205
Dong 13 Ponkan	37
Dongjiang Bendizao	16
Dongshizao Pummelo	207
Dream Navel Orange	181
Duweiwendanyou	208

E

Early-rippening Jiaogan	78
E-gan No.1 Ponkan	38
E-gan No.2	90
Eureka Lemon	245

F

Fallglo	122
Feicuiyou	209
Fengcai	153
Fengjie 72-1 Navel Orange	169
Fengwan Navel Orange	170
Fingered Citron	238
Frost Navel Orange	172
Frost Valencia Orange	138
Fuju	23
Fukumoto Navel Orange	171
Fulianju	18
Fuyouxuan Jiaogan	73

G

Gailiang Orange	139
Ganan No.1 Navel Orange	173
Gannan Zao Navel Orange	175
Gaocheng	102
Gonggan	68
Goutoucheng	199
Guanximiyou	210
Guicheng No.1	141
Guijingan No.2 Kumquat	258
Guijingan No.1 Kumquat	256

Guiju No.1	19
Guixiacheng No.1 Summer Orange	142
Guiyou No.1 Pummelo	212
Guoqing No.1	85

H

Hamlin Sweet Orange	143
Hanggan	20
Harehime Tangor	121
HB Pummelo	202
Heyang No.2 Ponkan	39
Himekoharu Tangor	129
Hiruka Tangor	99
Hongbaoshi Pummelo	213
Hongjin Tangor	103
Hongju	21
Hongpisuanju	55
Hongshigan	106
Hongyugan	107
Hongyun Xianggan Tangor	108
Huagan No.2 Ponkan	40
Huagan No.4 Ponkan	42
Huangguogan	70
Huayou No.2 Pummelo	215
Huazhou Orange	144
Huyou	234

I

Ichifumi Wase	88
Inaba Wase	82
Iyokan	128

J

Jiangan	26
Jiaogan	71
Jincheng	145
Jinchun Tangor	109

Index to Citrus Varieties in China (in English)

Jindan	260
Jindou	261
Jinhong Bingtangcheng	146
Jinhuang Tangor	111
Jinlanyou	216
Jinlegan Tangor	112
Jinqiu Shatangju	113
Jinxiangyou	217
Jinxiu Bingtangcheng	147
Jiuyueju	27
Juhong	218
Juxianglong Bingtangcheng	148
Juxiangyuan Sweet Orange	149

K

Kaixuangan	114
Kanpei	101
Kiyomi	119
Koenoakebono	83

L

Lane Late Navel Orange	179
Langfeng Navel Orange	178
Liangpingyou	219
Lianhong	94
Linghaoxuegan	165
Liucheng	151
Liuchengmiju	32
Longyuanzao	86
Luofu	262
Luowen	263

M

Manju	29
Mantouhong	61
Mashuiju	28
Meyer Lemon	242
Midknight Valencia Orange	156
Mingliucheng	154
Mingliutianju	30
Miyagawa Wase	91
Miyamoto Wase	84
Murcott	116

N

Nan No.3 Jiaogan	74
Nanfengmiju	31
Nanju	25
Nankan No.20	95
Nankou	117
Navelina Navel Orange	182
Newhall Navel Orange	184
Nianju	34
Nichinan No.1 Wase	87
Niurouhong Zhuju	62
Nova	118

O

Oita	81
Okitsu Wase	93
Olinda Valencia Orange	131
Omishima Navel Orange	168
Orah	125
Oroblanco	232
Ougan	79
Owari	96

P

Pengju	51
Pingshanyou	220
Ponkan	36

Q

Qianyang Bingtang Navel Orange	186
Qianyang Seedless Ponkan	43
Qingqiu Navel Orange	187

R

Red Limonia	244
Red Tangerine	21
Robertson Navel Orange	180
Rohde Red Valencia Orange	155
Rough Lemon	243
Ruanzhisuanju	56
Ruby Blood Orange	195
Ruby Pummelo	213
Ruju	52

S

Sangmayou	223
Sanhongmiyou	222
Sanhuhongju	63
Satsuma Mandarin	80
Seike Navel Orange	188
Shaohe Gonggan	69
Shaohe Nianju	35
Shatangju	53
Shatianyou	225
Shi 18 Ponkan	44
Shiranui	98
Shitouju	200
Sijiju	265
Sijipao	226
Skaggs Bonanza Navel Orange	185
Sour Pummelo	227
Sour Tangerine	55
Star Ruby	237
Suanju	55

T

Ta 59 Jiaogan	75
Tachima Wase	92
Tang Orange	157
Tangor No.1232	97
Tankan	71
Taoye Orange	158
Tarrocco Blood Orange	197
Trifoliate Orange	248

V

Valencia Orange	137

W

Wanbaiyou	228
Washington Sanguine Blood Orange	196
Washington Navel Orange	177
White Limonia	241
Wilking	57
Wuyueju	58

X

Xianfengcheng	159
Xiangshui Orange	160
Xiaoxuegan	166
Xin No.1 Jiaogan	76
Xinhui Sweet Orange	161
Xinshengxi No.3 Ponkan	45
Xuegan	163
Xupuchangxing Sweet Orange	162

Y

Yangguang No.1 Tangelo	127
Yangshanju	59
Yanxiwanlu Ponkan	46
Yongchun Ponkan	47
Yuanfeng Navel Orange	190
Yuefeng Jiaogan	77
Yuefengzaoju	33
Yueyingtianju	130
Yueyou Ponkan	48
Yuhuanyou	229
Yunning No.1 Lemon	247
Yura Wase	89

Z

Zaohong Navel Orange	192
Zaoju	60
Zaomi Ponkan	50
Zaoshu Jiaogan	78
Zaoxiangyou	230
Zaoyuwendanyou Pummelo	231
Zhonggan Micheng	167
Zhuhongju	64
Zhuju	61
Zhuluan	201
Zhushaju	66
Zijinchuntianju	67
Zongcheng Navel Orange	193